Brewing Engineering

Making Great Beer Through Applied Science

2nd Edition

Steven Deeds

BREWING ENGINEERING

© Copyright 2013 by Steven Deeds

All rights reserved. No portion of this book may be reproduced in any form without written permission of the publisher. Neither the authors, editor nor the publisher assume any responsibility for the use or misuse of information contained in this book.

Printed in the United States of America

10 9 8 7 6 5 4 3 2

ISBN-10: 1482540509
ISBN-13: 978-1482540505

Library of Congress Cataloging-in-Publication Data

    Deeds, Steven, 1977-
    Brewing Engineering : Making Great Beer Through Applied Science.
    p. cm.
    ISBN-13: 978-1482540505
    I. Brewing II. Title

Copy Editing: Jessica Kantrowitz

Dedicated to my daughter Abigail.

Who, with the printing of this book, I will be able to spend more time.

This book would not have been possible without

the loving kindness of my wife, Terri.

Nor would it have been nearly as extensive without

the many knowledgeable brewers

that have so freely shared their experiences and discoveries.

Forward

What is great beer?

The proof is in the pudding as they say, or the beer in this case. It doesn't matter how you get there, so much as that you have arrived. When evaluating a brew it's easy to assess the numbers: Did it attenuate well? Was the correct amount of yeast pitched? Was the temperature controlled adequately? It's easy to lose sight of the goal: what matters is how it tastes.

What makes great beer?

Good beer comes from understanding the process and the ingredients. Great beer takes everything one step higher. Great beer is made by consistent, repeatable processes. It comes from knowing what to change and learning how to change it. This is the foundation of experimental science, and to that extent, even artistic and creative brewers have a little bit of a scientist in them. It's from that perspective that this book is written.

Great beer is the result of diligence.

Table of Contents

Forward 4
Table of Contents 5
List of Tables 9
Recipe Formulation 11
    Bittersweet 12
    Recipe Scaling 15
    How to Make a Beer Recipe Clone 17
    Hop Chart 20
    Final Gravity 23
    Predicting Final Gravity 24
    Crystal Malt Fermentability 27
Water 29
    Basic Water Chemistry 30
    Reducing Chlorine and Chloramine 32
    Adjusting Mash pH 34
Mashing 37
    Brew in a Bag (BIAB) 38
    Mash Temperature and Thermometers 39
    Calibrating your Thermometer 41
    Mash Temperature Theory 44
    Measured Mash Temperature Effects 46
    Predicting Brewhouse Efficiency 48
    More Grain Isn't Always More Sugar 51
    Maximizing Your Mash 53
    Wort Sugars 55
    Hitting the Exact Original Gravity 58
More Than Malt 61

- Fruit the Easy Way 62
- How Much Hops 64

Growing Yeast 66
- Visualizing Growth 67
- Starter Calculators 69
- Viability vs. Vitality 70
- Measured Cell Growth 72
- Yeast Growth Rate 75
- Anaerobic and Aerobic Respiration 78
- Extrapolated Data 81
- Cell Growth as a Function of Sugar 87

Yeast Pitching 90
- Starter Cell Count Estimators 91
- Rehydrating Yeast 93
- How Many Cells in a Package 95
- Yeast Pitching Rates 97
- Reduce the Wort Instead of Increasing the Yeast. 98
- Do the Two Step 100
- Aeration 102
- Pitch Rates and Starters 104

Storing Yeast 108
- Yeast "Washing" 109
- Rinsing Yeast From a Fruit Beer 111
- Yeast Storage 112
- Refrigeration and Viability 114
- Yeast at Ambient 116
- ABV Effects on Yeast 118
- Fruit and Viability 121

Real Yeast Washing 123
Summary of Washing Techniques 124
Acid Washing 125
Fermentation 127
How Often to Check Gravity 128
Calculating Fermentation Time 130
How Long Does it Take to Carbonate? 132
Trub Loss 133
Fermentation Control 135
Swamp Cooler 136
Aquarium Heater 142
Lagering Outside 144
Packaging 146
Not Just Sanitized, but Sterilized! 147
Easy Priming Sugar 148
Matching ABV with Priming sugar. 150
Brew Lab Equipment 152
Hydrometer vs. Refractometer 153
Selecting a Microscope 156
Hemocytomeres 158
Brew Lab Technique 159
Overview 160
How to Make Your Own SRM Tester 162
Harvest Yeast From a bottle 164
Isolating Cells by Plating 167
Serial Dilution 171
Refractometer Corrections for Alcohol 173
ABV without OG 175

How Accurate is a Cell Count 180
Counting Yeast Cells to Assess Viability for Brewing 182
Trouble Shooting 187
Top Ten Reasons Why Your Final Gravity is Stuck 188
Top Ten ways to Restart Fermentation 190
Adjusting Flavor 192
Efficient Brewing 194
Great Beer with Little Time 195
Dry vs. Liquid Extracts 200
Bitter Without the Boil 201
Easy Pitching 204
Fast Brew Method 205
Adding Dimension 207
Mix to Taste 208
Recipes 209
Easy IPA 210
Brewing by the Numbers 212
Raspberry Hefeweizen 215
Raspberry Cream Ale 216
Hoegaarden Clone Recipe 219
Equations 221
Overview 222
Lautering 223
Wort Properties 223
Beer Properties 224
Attenuation 226
Yeast Growth 226
Index 227

List of Tables

Table 1 - Recipe Scaling Table 15
Table 2 - Low Alpha Hops 21
Table 3 - High Alpha Hops 22
Table 4 - Final Gravity Equivalent Sweetness 23
Table 5 - Apparent Attenuation of Brewing Sugars 24
Table 6 - Resulting pH from Neutralized Chlorine 32
Table 7 - Campden Tablet to Cancel Free Chlorine 33
Table 8 - Milliliter of 88% Lacic Acid to Adjust Mash pH 36
Table 9 - Teaspoons of 88% Lactic Acid to Adjust Mash pH 36
Table 10 - Efficiency From Water Volume and Grain Weight 49
Table 11 - Maximum Extracted Fermentables 52
Table 12 - Runnings From Tun Size and Grain Weight 53
Table 13 - Wort Sugar Composition 55
Table 14 - Stater Calulator Summary 69
Table 15 - Start Cell Count Over Time 77
Table 16 - Fermentation Matrix Results 82
Table 17 - Actual Attenuation from Innoculation Rate and OG 86
Table 18 - Cell Growth from Refractometer Measurements 89
Table 19 - Cell Counts of Various Packages 95
Table 20 - Ale Pitch Rate 97
Table 21 - Aeration 103
Table 22 - Starter Size from Packaged Yeast 106
Table 23 - Starter size for Slurry Yeast 107
Table 24 - Yeast Washing Cell Counts 111
Table 25 - Viability Over Time 115
Table 26 - Viability of Yeast as a Function of ABV 120
Table 27 - Post Fermentation Viability 122
Table 28 - Summary of Bacterial Contaminates 123
Table 29 - All Grain vs Extract 133
Table 30 - Swamp Cooler Thermal Coefficient 140

Table 31 - Water and Priming Sugar	151
Table 32 - IOR and Density	154
Table 33 - SRM Equivelents	163
Table 34 - Gelatin Plate Recipe	168
Table 35 - Serial Dilutions	172
Table 36 - Number of Dilutions	172
Table 37 - OG from Final Measurements	178
Table 38 - ABV Without Original Measurments	179
Table 39 - All Grain to Extract Conversion	198
Table 40 - Extract Mineral Concentration for a 1.060 Wort	199
Table 41 - Salt Adjustments for Style	199
Table 42 - Bittering Hop Tea	202
Table 43 - Dry Yeast Comparison	204
Table 44 - Batch to Sample Conversions	208
Table 45 - Beer Styles 1-9	212
Table 46 - Beer Styles 10-19	213
Table 47 - Beer Constitues Durring Normal Fermentation	222

RECIPE FORMULATION

1

Bittersweet

The balance of bitterness and sweetness of a beer is complicated to achieve. While these tastes are perceived by different receptors on the tongue, their careful combination creates a balance. Too much sweetness without bitterness will be perceived as cloying; likewise too much bitterness without sweetness will taste soapy. Yet these two must be balanced perfectly on the teeter-tottering see-saw that is beer. The two together will be better than either on its own.

Dextrin Malts

Adding high dextrin malts like caramel or crystal will leave the beer with a higher final gravity.[1] Steeping crystal malt after the mash will produce 25% less fermentable sugars. This not only gives the beer a thicker mouth feel, but also adds to the final sweetness of the beer. High final gravity can also be achieved by mashing at higher temperatures. For every degree Fahrenheit above 155 the final attenuation of the beer is lowered about 2 percent.[2] In general 1%-2% is normal to add color to a beer. Five percent will add noticeable sweetness and caramel taste. Ten percent can easily be too much if not balanced with hops, or by keeping the chloride ions low.

IBUs

On the bitterness side, the most apparent factor is the hop level measured in IBUs. The bitterness from hops can easily be adjusted during the boil of the brewing process by changing boil time and amount of the hops. After the wort has cooled, the gravity sample can be tasted and dry hops can be added to further adjust the flavor. But because the wort will be high in sugar content at this time, it may be

---

[1] http://www.homebrewtalk.com/f128/testing-fermentability-crystal-malt-208361/index11.html
[2] http://braukaiser.com/wiki/index.php?title=Effects_of_mash_parameters_ on_fermentability_and_efficiency_in_single_infusion_mashing

hard to judge if more bitterness is needed. Keeping the IBUs around 20 will make most people happy although light beer drinkers may prefer less than eight. If the water contains a low amount of sulfates, the IBU value can be higher without tasting harsh.

Alcohol

After fermentation is complete most of the sugar will have been converted to alcohol. Higher alcohol content will mask some of the sweetness of the beer. Have you ever tasted a cocktail mix without the alcohol? They are extremely sweet, yet when the alcohol is added they taste much less sweet. The ABV to sweetness balance works itself out pretty well in beer because the alcohol content is directly proportional to the amount of sugar that is converted. So bigger beers have more final sugar, but are balanced by more alcohol.

Salts

Before bottling, the bitter sweet balance can be adjusted using salts. It's very easy to overdo it here. Remember these are called brewing SALTS, and if overused, you will have beer that tastes … well… salty! One-half teaspoon of salt in a five-gallon batch is normally plenty. Sulfates (SO_4^{-2}) will enhance the bitterness, while chloride ions (Cl^-) and sodium ions (Na^+) will enhance the sweetness. Sodium can often cause a metallic taste in the beer so common table salt (NaCl) is not often used in beer. There are a number of good sources online that can be found by searching for "brewing water chemistry."

CO_2

The final adjustment to balance is carbonation. The higher the volume of CO_2 dissolved in the beer the more bitter the beer will taste. If you compare a flat soda to a carbonated soda the difference in taste is very apparent. Although CO_2 content does play into the bitter/sweet balance it is likely the least noticeable factor. Considering it for taste should be secondary to mouth feel and style.

Yeast

Although the flavor of the yeast itself is not often present in the beer, the flavors that the yeast produces often are. Yeast produces esters and phenolic compounds that provide a range of distinct flavors. Belgium yeasts produce spicy notes; Bavarian yeasts produce clove or banana flavors; English yeasts are known for their fruitiness. These flavors will be masked by high alcohol, bitterness and hop flavor.

Finding this balance can take time. If you are starting a recipe from scratch it is easy to miss your mark by a mile. If you start with a reliable recipe you are bound to get good results.

Recipe Scaling

A recipe can be converted from one size to another using a number of means, but often a handy table is easier to use and plenty accurate.

		kg		DME	LME		lbs		Grain		
1.5	2.5	3.175	4	5	5.5	6	7	8	9	10	11	12
0.30	0.50	0.64	0.80	1.00	1.10	1.20	1.40	1.60	1.80	2.00	2.20	2.40
0.33	0.55	0.70	0.88	1.10	1.21	1.32	1.54	1.76	1.98	2.20	2.42	2.64
0.36	0.61	0.77	0.97	1.21	1.33	1.45	1.69	1.94	2.18	2.42	2.66	2.90
0.40	0.67	0.85	1.06	1.33	1.46	1.60	1.86	2.13	2.40	2.66	2.93	3.19
0.44	0.73	0.93	1.17	1.46	1.61	1.76	2.05	2.34	2.64	2.93	3.22	3.51
0.48	0.81	1.02	1.29	1.61	1.77	1.93	2.25	2.58	2.90	3.22	3.54	3.87
0.53	0.89	1.12	1.42	1.77	1.95	2.13	2.48	2.83	3.19	3.54	3.90	4.25
0.58	0.97	1.24	1.56	1.95	2.14	2.34	2.73	3.12	3.51	3.90	4.29	4.68
0.64	1.07	1.36	1.71	2.14	2.36	2.57	3.00	3.43	3.86	4.29	4.72	5.14
0.71	1.18	1.50	1.89	2.36	2.59	2.83	3.30	3.77	4.24	4.72	5.19	5.66
0.78	1.30	1.65	2.07	2.59	2.85	3.11	3.63	4.15	4.67	5.19	5.71	6.22
0.86	1.43	1.81	2.28	2.85	3.14	3.42	3.99	4.56	5.14	5.71	6.28	6.85
0.94	1.57	1.99	2.51	3.14	3.45	3.77	4.39	5.02	5.65	6.28	6.90	7.53
1.04	1.73	2.19	2.76	3.45	3.80	4.14	4.83	5.52	6.21	6.90	7.59	8.29
1.14	1.90	2.41	3.04	3.80	4.18	4.56	5.32	6.08	6.84	7.59	8.35	9.11
1.25	2.09	2.65	3.34	4.18	4.59	5.01	5.85	6.68	7.52	8.35	9.19	10.03
1.38	2.30	2.92	3.68	4.59	5.05	5.51	6.43	7.35	8.27	9.19	10.11	11.03
1.52	2.53	3.21	4.04	5.05	5.56	6.07	7.08	8.09	9.10	10.11	11.12	12.13
1.67	2.78	3.53	4.45	5.56	6.12	6.67	7.78	8.90	10.01	11.12	12.23	13.34
1.83	3.06	3.88	4.89	6.12	6.73	7.34	8.56	9.79	11.01	12.23	13.45	14.68
2.02	3.36	4.27	5.38	6.73	7.40	8.07	9.42	10.76	12.11	13.45	14.80	16.15
2.22	3.70	4.70	5.92	7.40	8.14	8.88	10.36	11.84	13.32	14.80	16.28	17.76
2.44	4.07	5.17	6.51	8.14	8.95	9.77	11.40	13.02	14.65	16.28	17.91	19.54
2.69	4.48	5.69	7.16	8.95	9.85	10.75	12.54	14.33	16.12	17.91	19.70	21.49
2.95	4.92	6.25	7.88	9.85	10.83	11.82	13.79	15.76	17.73	19.70	21.67	23.64
3.5	5.9	7.5	9.5	11.82	13.0	14.2	16.5	18.9	21.3	23.6	26.0	28.4

Table 1 - Recipe Scaling Table

Find the column for the batch size of the recipe, and the column for the scaled recipe. This table can also be used to convert from DME, LME and grain as well as pounds to kilograms. If the number is not on the table the decimal point will need to be moved. The resolution of this table is designed to provide 95% accuracy.

Example 1: *Convert 2 lbs of grain in a 5-gallon batch to a 3-gallon batch.*

Brewing Engineering

Find the column labed 5 (The fifth from the left). Find the closest value to 2 in that column (1.95). Find the column labled 4 (the fourth from the left) read across the same row that was found in the first column to the secound column (1.56). To replace 2 lbs of grain in a 5-gallon batch use 1.56 lbs of grain in a 4-gallon batch.

Example 2: *Converting a recipie that calls for 5 lbs of grain to DME.*

Find the column labeld Grain (fourth from the right). Find the closest value to 5 in that column (5.14). Find the column labled DME (fifth from the left). Read across from the cell found in the first column to the secound coulmn (2.85). To replace 5 lbs of Grain use about 2.75 lbs of DME.

Example 3: *When the numbers are off the scale. (1 lb of grain to LME)*

Find the Column labled Grain (fourth from the right) Note that 1 is less than the lowest value on the table. Instead move the decimal point one place to the right to make 10. The closest value is 10.01. Read to the left in the same row to the column for LME. This value is 6.12. Because we multiplied by ten, we must now divide by ten. One lb of grain is equivelent to 0.6 lbs of LME.

How to Make a Beer Recipe Clone

When cloning a beer there are numerous places that you can gather information.

1) The brewer's website.
2) Your own measurements and experience.
3) The BJCP style guide.
4) Sites that rate beers.
5) Forums and other online recipes.

The brewer's website

If you start with the facts then the assumptions and estimations that come up later in the process will not be as influential in your recipe. Most brewers are very proud of the product that they make and will tell you quite a bit about what makes their beer different, which almost always dives into some of the process and ingredients used. The Hoegaarden website has several important clues to the recipe. Hoegaarden is described as a classic wit beer with all of the ingredients grown in Belgium. Belgium pale malt is therefore likely the main base malt. The mash is 2 hours and 30 minutes and during this time the temperature is gradually brought to 75°C (167°F). If we assume a linear rise from room temperature, the protein rest (113°F to 133°F) is about 30 minutes, the Saccharification rest is about one hour, and the average temperature is 150°F. This is a pretty standard mash for a light bodied beer. As an alternative to converting this to a step mash, the same rate of 1.5 minutes per degree could be used to achieve a ramping mash. The boil time is one hour which is very standard. Citrus and coriander aroma and taste are mentioned.

Measurements

The ABV is stated as 4.9% from which the OG can be calculated if the FG is known. By degassing the beer, the FG can be measured. A sample tested indicates that the final gravity is 1.014. Knowing the FG and the ABV we can calculate the OG.

$$OG = \frac{ABV}{131} + FG => OG = 1.051$$

Alternatively, the gravity can also be calculated from the ABV and the number of calories. From the OG and FG the attenuation of the yeast can be calculated.

$$Attenuation = 1 - \frac{FG}{OG} => Attenuation = 72.5\%$$

The attenuation is lower than typical ale yeasts; however this is fairly typical for hefeweizen and other yeasts commonly used for wheat beers. Using the Whitbread Ale yeast from White Labs or Wyeast matches the attenuation, but Belgian Wit Ale seems to be more popular for this style.

Color can be compared to a BJCP color card to get an idea of the SRM. Or if you are really geeky you can measure it using a blue LED with 430nm wave length and matching photo detector. (Also see the section on measuring SRM.) Using the BJCP card Hoegaarden looks like a three or four.

Experience

When tasting Hoegaarden the citrus flavor jumps right out. There is also a fermented fruit flavor which makes me think they are using more than just the rind. There may be some pulp in this beer. The hop flavor is very subtle, but what is there has a distinct citrus flavor and very little bitterness. There are likely no bittering hops, and only flavor or aroma additions. My nose and taste buds say citrus, and the brewer's website says Belgium. Glacier matches these criteria so it should work well.

The BJCP style guide

We're lucky because the first example listed in the BJCP style guide for category 16A is...you guessed it, Hoegaarden Wit. Even if the beer you are cloning isn't listed, using the ingredients for the appropriate style will still be of great value. From this information we gather the grain bill should be 50% base barley malt and 50% wheat malt. The other additions that are mentioned are the same that we saw on the Hoegaarden web site -- orange peel and coriander. Looking at the other spices mentioned, and after a second taste of the beer, there may be some cumin in the recipe as well. Hoegaarden doesn't exactly fit into the style's "vital statistics" with the FG at 1.014 where the style guide shows the high end of 1.012. The IBUs for the style are 10-20. My taste buds tell me that it is at the lower end of this range.

Sites that rate beer

The two big ones, Beer Advocate and Rate Beer, provide access to hundreds of different tasting notes on this product. Perusing them confirms the ingredients that have been selected thus far. It's also worth noting that the beer is described as sweet by many people. With the low mash temperature the beer would typically be dry, not sweet. Therefore there is likely some dextrin or other sugars remaining unfermented. This could be from malt high in dextrin such as crystal malt, or from Carapils which is also high in long chain sugars. The color of Hoegaarden is very light, so Carapils is more likely. Another way to increase residual sweetness is to select yeast that does not attenuate well, and to ferment at colder temperatures.

Forums and other online recipes

Between Homebrewtalk, Beer Smith, and a plethora of blogs and other sites the available information is overwhelming. Choosing three to five recipes to compare will provide adequate information.

Hop Chart

Selecting hops for a recipe can be a daunting task. With literally more than one hundred varieties where do you start? It would take several experimental batches of beer and a pretty substantial monetary investment to try even what is available at your local home brew store. A reasonable place to start may be with recipes created by sources you trust. Another way would be to break down the data available.

There is an enormous amount of flexibility here, but this should help you narrow down the selection to a few tried and true examples. Once you've settled on a style of beer you want to brew, the hops can be derived from what is appropriate for that style.

1. Choose between high alpha hops for bittering, and low alpha hops for aroma and flavor.
2. Distinguish between Ale and Lager.
3. Select the region based on proximity to the styles origin.
4. Select an appropriate flavor.

Making Great Beer Through Applied Science

Saaz	Tett	Hall	Libr	Fugg	Will	Gold	Styr	Mt.H	Casc	
4.0%	4.5%	4.5%	4.5%	4.8%	4.8%	5.0%	5.3%	5.5%	5.8%	Alpha Acid
0.6%	0.6%	0.8%	0.9%	1.0%	1.2%	0.8%	0.8%	1.1%	1.1%	Flavor & Aroma
	X			X	X	X	X		X	Ale
X	X	X	X				X	X		Lager
		X	X				X	X		American
				X	X	X	X			English
X	X	X								German
					X		X			Bitter
	X	X						X		Bock
				X	X	X				ESB
									X	IPA
				X	X				X	Pale Ale
X		X	X					X		Pilsner
					X				X	Porter
					X					Stout
X		X	X					X		Wheat

Saaz	Tett	Hall	Libr	Fugg	Will	Gold	Styr	Mt.H	Casc	
X	X	X		X			X		X	Spicy
	X									Herbal
X			X	X						Earthy
		X		X	X				X	Floral
			X	X					X	Fruity
									X	Citrusy

Sazz = Sazz   Tett = Tettnang   Libr = Liberty
Fugg = Fuggles   Will = Willamette   Gold = Goldings
Styr = Styrian   Mt. H = Mt. Hood   Casc = Cascade

Table 2 - Low Alpha Hops

BREWING ENGINEERING

Clus	Chal	Perl	NB	Cent	Gale	Chin	Nugg	Magn	
7.2%	7.5%	7.8%	8.0%	9.8%	12%	12%	13%	14%	Alpha Acid
0.6%	1.4%	0.8%	1.7%	1.9%	1.0%	2.0%	1.9%	1.7%	Flavor & Aroma
X	X	X	X	X	X	X	X		Ale
X								X	Lager
X			X	X		X			American
	X		X		X		X		English
		X						X	German
X	X		X		X		X		Bitter
									Bock
	X		X		X		X		ESB
						X			IPA
		X	X			X			Pale Ale
								X	Pilsner
	X	X	X		X		X		Porter
	X				X	X	X	X	Stout
			X						Wheat

Clus	Chal	Perl	NB	Cent	Gale	Chin	Nugg	Magn	
X	X	X				X	X		Spicy
							X		Herbal
			X						Floral
			X	X					Citrusy

Clus = Cluster   Chal = Challenger   NB = Northern Brewer
Cent = Centennial   Gale = Galena   Chin = Chinook
Nugg = Nugget   Magn = Magnum

Table 3 - High Alpha Hops

Final Gravity

After fermentation has completed, nearly all of the simple sugars such as sucrose and glucose have been converted to alcohol. Most of the maltose has also been converted. What remain are mostly long chain sugars, such as dextrin sugars and maltotriose. Dextrin sugars are relatively tasteless, but Maltotriose can be broken down by enzymes in your mouth to form glucose.[3]

Because some of the sugars remaining in beer cannot be tasted, drinking beer and noting the level of sweetness and final gravity is likely the best way to gauge how sweet your beer may turn out. However these comparisons may give you a new way to think about residual sweetness.

How much sugar do you put in your tea or coffee?

The alcohol and hop level of a beer balance out the residual sugar the same way bitterness in tea or coffee can balance the sugar that you might add. Coffee is a closer comparison for highly hopped beer as they are both quite bitter without a little sugar, and tea might be a good comparison for a beer with low alcohol and hops.

That Said, I like my coffee without any sugar, but prefer a Belgian Quad to a Double IPA. This might give you a new way to think about final gravity, but it's certainly not a perfect comparison.

Final Gravity		Sugar in 1 Cup of Water
1.008	2°P	1 tsp
1.015	4°P	2 tsp
1.023	6°P	3 tsp
1.031	8°P	4 tsp
1.039	10°P	5 tsp

Table 4 - Final Gravity Equivalent Sweetness

---

[3] http://en.wikipedia.org/wiki/Maltotriose

Predicting Final Gravity

The attenuation of beer styles listed in the BJCP style guide ranges from 63% for a 9E Strong Scotch Ale up to 97% for the 1A Light American Lager!

Having control over your attenuation is necessary for these styles and actually fairly simple to accomplish.

The primary driver of attenuation is the sugar constitute of the wort. The yeast strain, pitch rate, and temperature affect attenuation to a lesser degree

All beer yeasts have the ability to ferment maltose, but there is some variation in how much they leave behind. Attenuation of given yeast is essentially its ability to ferment more complex sugars. Today's brewing yeast provides a range from about 65% for the English Ale Yeasts to 80% for Belgian Saison strains. This gives the brewer almost enough room to cover the whole spectrum.

In addition to yeast strain selection fermentability can be manipulated with maltodextrin to increase the final gravity or sucrose, to lower the final gravity. To determine just how much these alternative sugars affect attenuation an experiment was conducted in triplicate on nine different ratios of sugars. All fermentations used 1 million cells per ml per degree Plato. The starting gravity of each was 9°P which is a specific gravity of 1.036.

% Sucrose	0%	0%	0%	0%	0%	25%	50%	75%	100%
% Mato-Dextrin	100%	75%	50%	25%	0%	0%	0%	0%	0%
% DME	0%	25%	50%	75%	100%	75%	50%	25%	0%
Test 1 Attenuation	0%	12%	31%	49%	67%	80%	84%	91%	99%
Test 2 Attenuation	0%	15%	37%	51%	67%	75%	85%	93%	102%
Test 3 Attenuation	0%	15%	32%	50%	67%	75%	86%	94%	96%
Average	0%	14%	33%	50%	67%	76%	85%	93%	99%

Table 5 - Apparent Attenuation of Brewing Sugars

The results were predictable, and can be easily applied to recipe formulation. The apparent attenuation of the dry malt extract was 67% which is the middle of the expected range for the S04 yeast that was used. Maltodextrin was not converted at all. Sucrose had an apparent attenuation of about 100%.

When brewing all-grain, the mash temperature has one of the largest impacts on fermentability of the wort. Higher temperatures will break down the beta enzymes and also speed the rate at which the alpha enzymes work. This leaves more complex sugars in the wort which are less fermentable. Mash times longer than 60 minutes will also lead to more attenuation of the sugar during fermentation as the enzymes have more time to break down the complex sugars further. The results of fairly extensive experiments, as well as my own experience, indicate that each degree above 151 will yield 1.5% less attenuation. (See the section on Mashing for more information.)

For practical purposes it can be assumed that crystal malt ferments nearly as well as base malt. This is especially true when the crystal malt is added to the mash. Here the enzymes will convert some of the complex sugars that were otherwise unconverted.

From these factors, the final gravity can be calculated. Let's use a simplified recipe for a poorly designed milk stout as an example.

5-gallon batch.

7 lbs 2 row.
1 lb sucrose.
1 lb lactose.
WLP004 (72% attenuation)
154 mash temperature
75% mash efficiency

The OG contribution of the 2 row is:

7 lbs * 37 ppg * 75% / 5 gallons = 38.85 gravity points

The OG contribution from the sucrose is:

1 lbs * 46 ppg / 5 gallons = 9.2 gravity points

The OG contribution from the lactose is:

1 lb * 43 ppg / 5 gallons = 8.6

The original gravity is therefore:

38.85 + 9.2 + 8.6 = 56.65 gravity points or a specific gravity of 1.057

The final gravity of the 2 row is affected by both the yeast and the mash temperature. The temperature is 2 degrees above 152, so that means the malt will be 4% less fermentable.

72% - 4% = 68% attenuation of the malt contribution.

The FG contribution of the 2 row is:

38.85 * (1 - 68%) = 12.4

The FG contribution of the sucrose is zero because it will ferment out dry, and the contribution of the lactose is 8.6 because it will not ferment at all.

The final gravity is therefore:

12.4 + 0 + 8.6 = 21 gravity points or 1.021 specific gravity.

The apparent attenuation is much lower than what is specified for the yeast alone.

1 - (21/57) = 63%

Even with the sucrose in the recipe to dry it out, the final gravity may be high for some people. To repair this recipe, yeast that attenuates more could be selected, the mash temperature could be lowered, or the lactose could be reduced.

Crystal Malt Fermentability

For the most part you are safe assuming that crystal malt will ferment the same as your base malt, especially considering that the crystal will be a relatively small portion of your grain bill. Additions of lactose and sucrose will affect final gravity much more, but when it really comes down to fine-tuning your recipe calculations, you might want to consider how fermentable crystal malt is. It could change your final gravity by a couple of points.

Fairly extensive tests of crystal malt fermentability were conducted over the course of several months.[4] The results of these tests are a look into how fermentable crystal malt is when mashed with base malt.

Average attenuation

    80% for 2 row
    77% for 50% C10, 50% 2 row
    70% for 50% C40, 50% 2 row
    67% for 50% C120, 50% 2 row

If we assume that the 2 row ferments the same regardless of being mashed with the crystal the 100% crystal fermentability numbers can be derived.

    74% for C10
    60% for C40
    54% for C120

If you really want to fine-tune your recipes you can use the differences in fermentability to adjust your final gravity calculation. These would be:

---

[4] http://www.homebrewtalk.com/f128/testing-fermentability-crystal-malt-208361/index11.html

Brewing Engineering

-6% for C10
-20% for C40
-26% for C120

For example, say you were making a dark beer with 9 pounds of 2 row, and 1 pound of crystal. Based on your mash temperature and type of yeast you can determine your base malt fermentability as described in the "Final Gravity in Recipe Formulation" section. For this example we will use 75% base malt fermentability.

OG would be (note that mash efficiency has been rolled into the ppg numbers)
9 lbs of 20 row * 24 ppg = 216 GU
1 lb of C120 * 23 ppg = 23 GU
216+23 / 5 = 48 GP or a specific gravity of 1.048

FG would be as follows
216 GU * (1-75%) = 54 GU
23 GU * (1-75%-26%) = 4 GU
54+4 / 5 gallons = 11.6 GP or a specific gravity of 1.012

If the fermentability of the crystal was not considered then the contribution would be 6 GU instead of 4. Because this change is very small, the FG of the five-gallon batch would still be 1.012

So, as you can see, most of the time the change in attenuation caused by crystal malt is insignificant.

WATER

2

Basic Water Chemistry

Water ion composition is one of the more subtle ways to fine-tune a beer, but if used in excess it can also make your beer unpalatable.

Keeping it simple will keep you out of trouble. In most of the United States tap water has a fair amount of sulfates, and too much of both sodium and chlorine for most beers. In addition, it tends to have very little calcium. Mash enzymes need calcium to work effectively, and it is a vital nutrient for yeast. The challenge with this water is boosting the calcium level without overloading the other minerals.

Boston tap water (used for Sam Adams Beer) has the following mineral profile:

Ca 4 ppm
Mg 1 ppm
Na 32 ppm
Cl 23 ppm
SO_4 6ppm
$CaCO_3$ 41ppm

The overall hardness (or total dissolved solids) is low which allows it to be built up. If your water is hard (above 120 ppm $CaCO_3$) then you will probably want to use reverse osmosis or distilled water.

CaCl (Calcium Chloride) will make beer taste sweeter. For making a beer that has some malty sweetness to it, add ½ a tsp of Calcium Chloride to your grains for a 5-gallon batch of beer. This will add 49 ppm calcium and 86 ppm chlorine.

$CaSO_4$ (Calcium Sulfate or Gypsum) will bring out the hop bitterness. For pilsners, lagers and light ales add ¼ tsp Gypsum. For a middle of the road ale, add ½ tsp. For Stouts, add ¾ tsp and for IPAs and other hoppy beers, add 1 tsp.

CaCO$_3$ (Calcium Carbonate or Chalk) for a dark beer that you want to add some mineral water qualities to add up to 1 tsp of Calcium Carbonate. Unlike the other brewing salts, chalk will raise the pH of the mash water, however most of the time the pH needs to be lowered in the mash to achieve optimal pH levels.

You can make it much more complicated than this if you want, but these basics should give you a starting point for fine-tuning your beer.

Reducing Chlorine and Chloramine

The EPA limit for Cl- in water is 4 ppm. Hydrogen peroxide, 2(HO), will react almost immediately with free chlorine (Cl-) in water forming hydrochloric acid (HCl), a strong acid. Even at 4 ppm the pH of the solution can be quite low. If using hydrogen peroxide to reduce the chlorine in the water you can expect the resulting pH to be very low.

ppm of Cl-	Molar Solution	pH
0.5	1.38E-05	4.9
1	2.75E-05	4.6
1.5	4.13E-05	4.4
2	5.50E-05	4.3
2.5	6.88E-05	4.2
3	8.25E-05	4.1
3.5	9.63E-05	4.0
4	1.10E-04	4.0

Table 6 - Resulting pH from Neutralized Chlorine

It's common for water treatment facilities to use chloramine[5] in place of chlorine because it does not break down as readily. The downside for home brewers is that the reaction with 2(HO) is also much slower. Campden tablets, potassium or sodium metabisulfite, work well in this case.[6]

How much?

Take a look at your water report. If "Free Chlorine" is listed then 2(HO) will work. It only takes about 1 ml (1/4tsp) to cancel all of the free chlorine in five gallons of water.

For Chloramine, you may see a listing for "Total Chlorine" and a zero for "Free Chlorine" This indicates that all of the Chlorine is from Chloramine. You may see dosage instructions of one tablet for each

---

[5] http://en.wikipedia.org/wiki/Chloramine
[6] http://en.wikipedia.org/wiki/Campden_tablets

gallon. This is the level required to inhibit yeast growth, and is used in wine and cider making to stop fermentation. For reducing chlorine, the dosage is much lower. Each Campden tablet will reduce 3 ppm of Chloramine in 20 gallons of water. If an insufficient amount of the Campden tablet is used there will be remaining Chlorine in the water. If too much of the tablet is used there will be sulfur dioxide remaining. Either way is not great, but generally more sulfur dioxide is better than more chlorine. The sulfur aromas seem to disperse more readily than the chlorine. Figuring out the required amount is something that you will likely only have to do once. Because the Campden tablet addition will be treating all of the water used for your brew, you'll want to use the total amount of water even if the Campden tablet is just added to the boil. You'll just need to scale the recommended dosage based on the amount of water you will use to make the beer and the amount of total chlorine in your water.

W = Brewing water in gallons

C = total Chlorine in ppm

$$Amount\ of\ Campden\ tablet = \frac{W \times C}{60}$$

For example, here in Boston the water has 2.9 ppm total chlorine. For a typical brew 6 gallons of water are used to make a five-gallon batch. (2.9*6/60=0.29) For a typical brew day I'll use between a quarter and a third of a tablet depending on how the cookie crumbles, or tablet in this case.

Cl⁻	9	11	15	19	21	23	26	30	34	38	42	liters
ppm	2.5	3	4	5	5.5	6	7	8	9	10	11	gal
1.00	0	0	0	0	0	0	1/9	1/7	1/7	1/6	1/5	
2.00	0	0	1/7	1/6	1/5	1/5	1/4	1/4	2/7	1/3	3/8	
3.00	1/8	1/7	1/5	1/4	2/7	2/7	1/3	2/5	4/9	1/2	5/9	
4.00	1/6	1/5	1/4	1/3	3/8	2/5	1/2	1/2	3/5	2/3	3/4	

Table 7 - Campden Tablet to Cancel Free Chlorine

Adjusting Mash pH

It's hard to find any good rules of thumb out there for how much lactic acid to add to the mash to adjust the pH, and there's good reason for that. The starting mash pH is very dependent on a number of complicated factors and interactions. The main driver of the starting pH is the pH of the grain itself and the dissolved minerals in the water interacting with the grain. These will stabilize at some pH value when you have thoroughly stirred in your grains. How this all works is... complicated, and you really don't need to understand it to brew great beer. The fact is you really aren't going to know what your pH is until you get to dough in.

...but there is an easy way to figure out how much acid to add after dough in!

Once you have mashed in you can then measure and adjust the pH. While the starting pH is hard to determine, the strength of the buffering agent is not. Because there are so many more dissolved particles of grain in the water than dissolved particles of minerals, the buffer power of the mash is almost entirely driven by the amount of grain that you have added to your water. The more grain added to the water, the stronger the buffer. The stronger the buffer, the more resistant to change the pH will be. Because it is the grain driving the strength of the buffer, the pH change of a volume of acid will not be significantly affected by the amount of water in your mash-tun. Meaning that whether you have 2 gallons or 20 gallons of water, it will not affect how much 1 mL of acid changes the pH. The fact that the mash pH is not related to the volume of water may seem very strange, so let me explain further. pH is a driven by the ratio of positive ions to negative ions. The water itself has very few ions, so while it does have a pH value, it is a very weak buffer. The grain, however, has an enormous amount of particles that dissolve into the solution, meaning thousands of times more ions than the minerals in the water.

Therefore, the ratio of ions is driven by the grain, not the water. To beat a dead horse, because this was such a light bulb moment for me, let's use some numbers to explain it one more time. If the water had equal number of ions, five positive and five negative, (5+, 5-) the ratio would be one to one (1:1). If we add grain with ten thousand positive ions, and five thousand negative ions (10,000+, 5,000-) the ratio is now ten thousand and five to five thousand and five, (10,005:5,005) or about two to one (2:1) which is roughly the same as the grain alone.

Using the EZ-Water calculator,[7] and several brewing experiments, this simple equation was derived:

$mL\ of\ lactic\ acid\ needed\ =\ change\ in\ pH\ \times\ weight\ of\ grain\ in\ lbs$

For 10 lbs of grain you would need 1 ml for every 0.1 pH point.
For 5 lbs of grain you would need 0.5 ml for every 0.1 pH point

This is for 88% lactic acid solution which seems to be pretty common. To make measuring simple without a pipette use a normal eye dropper. It will hold about 1 mL and each drop is about 0.1 ml

---

[7] http://www.ezwatercalculator.com/

Hot Mash pH	\multicolumn{12}{c}{Pounds of Grain}
	3	4	5	6	7	8	9	10	11	12	13	14
6.5	3.9	5.2	6.5	7.8	9.1	10.4	11.7	13.0	14.3	15.6	16.9	18.2
6.4	3.6	4.8	6.0	7.2	8.4	9.6	10.8	12.0	13.2	14.4	15.6	16.8
6.3	3.3	4.4	5.5	6.6	7.7	8.8	9.9	11.0	12.1	13.2	14.3	15.4
6.2	3.0	4.0	5.0	6.0	7.0	8.0	9.0	10.0	11.0	12.0	13.0	14.0
6.1	2.7	3.6	4.5	5.4	6.3	7.2	8.1	9.0	9.9	10.8	11.7	12.6
6.0	2.4	3.2	4.0	4.8	5.6	6.4	7.2	8.0	8.8	9.6	10.4	11.2
5.9	2.1	2.8	3.5	4.2	4.9	5.6	6.3	7.0	7.7	8.4	9.1	9.8
5.8	1.8	2.4	3.0	3.6	4.2	4.8	5.4	6.0	6.6	7.2	7.8	8.4
5.7	1.5	2.0	2.5	3.0	3.5	4.0	4.5	5.0	5.5	6.0	6.5	7.0
5.6	1.2	1.6	2.0	2.4	2.8	3.2	3.6	4.0	4.4	4.8	5.2	5.6
5.5	0.9	1.2	1.5	1.8	2.1	2.4	2.7	3.0	3.3	3.6	3.9	4.2
5.4	0.6	0.8	1.0	1.2	1.4	1.6	1.8	2.0	2.2	2.4	2.6	2.8
5.3	0.3	0.4	0.5	0.6	0.7	0.8	0.9	1.0	1.1	1.2	1.3	1.4
	1.4	1.8	2.3	2.7	3.2	3.6	4.1	4.5	5.0	5.4	5.9	6.4
	\multicolumn{12}{c}{kg of Grain}

Table 8 - Milliliter of 88% Lactic Acid to Adjust Mash pH

Hot Mash pH	\multicolumn{12}{c}{Pounds of Grain}
	3	4	5	6	7	8	9	10	11	12	13	14
6.5	1/2	3/4	1	1 1/4	1 1/2	1 3/4	2	2 1/2	2 3/4	1 Tbls	1 Tbls	1 Tbls
6.4	1/2	3/4	1	1 1/4	1 1/2	1 3/4	2	2 1/4	2 1/2	2 3/4	1 Tbls	1 Tbls
6.3	1/2	3/4	3/4	1	1 1/4	1 1/2	1 3/4	2	2 1/4	2 1/2	2 3/4	2 3/4
6.2	1/4	1/2	3/4	1	1 1/4	1 1/4	1 1/2	1 3/4	2	2 1/4	2 1/2	2 1/2
6.1	1/4	1/2	3/4	3/4	1	1 1/4	1 1/2	1 1/2	1 3/4	2	2	2 1/4
6.0	1/4	1/2	1/2	3/4	1	1	1 1/4	1 1/4	1 1/2	1 3/4	1 3/4	2
5.9	1/4	1/4	1/2	1/2	3/4	1	1	1 1/4	1 1/4	1 1/2	1 1/2	1 3/4
5.8	1/8	1/4	1/4	1/2	1/2	3/4	3/4	1	1	1 1/4	1 1/4	1 1/2
5.7	1/8	1/4	1/4	1/4	1/2	1/2	3/4	3/4	3/4	1	1	1 1/4
5.6	1/8	1/8	1/4	1/4	1/4	1/2	1/2	1/2	3/4	3/4	3/4	1
5.5	1/8	1/8	1/8	1/8	1/4	1/4	1/4	1/4	1/2	1/2	1/2	1/2
5.4	1/8	1/8	1/8	1/8	1/8	1/8	1/8	1/4	1/4	1/4	1/4	1/4
5.3		1/8	1/8	1/8	1/8	1/8	1/8	1/8	1/8	1/8	1/8	1/8
	1.4	1.8	2.3	2.7	3.2	3.6	4.1	4.5	5.0	5.4	5.9	6.4
	\multicolumn{12}{c}{kg of Grain}

Table 9 - Teaspoons of 88% Lactic Acid to Adjust Mash pH

MASHING

3

Brew in a Bag (BIAB)

This is by far the easiest lautering technique I have tried! If you are thinking about going all grain this is by far the easiest way to do it. It's so clean! No transferring of hot wort. No tinkering with temperatures by calculating volumes to arrive at your Saccharification temperature. During the three step mash it was a breeze to hit all my temperatures spot on.

And if you don't know someone with a sewing machine, 5 gallon paint strainer bags work great.

Some may recommend squeezing the bag to remove more of the liquid, but I have found that just allowing it to drip yields nearly as much runnings. The bag can be set on a cooling rack to drip into the kettle serving as both the mash-tun and the boiling kettle. When the run-off is nearly complete, the volume can be measured and sparge water added to make up the difference that will be needed for the pre-boil volume.

Whether you are using BIAB or a mash tun your efficiency will be driven by water volumes and not the technique that is employed

The amount of grain is limited by the size of the kettle. With a 4 gallon pot the most that can be put in it is about 10 lbs of grain in 10 qts of water. Meaning, a five gallon stout recipe is simply not going to fit. You'll have to scale back your Guinness clone to 3 gallons or so. At 75% efficiency (which is normal for BIAB) and using 37 GP grain, that puts you at 277.5 Gravity points per gallon. So for a three-gallon batch the OG is 1.092, which is pretty good, but a five-gallon batch is 1.056 which would yield a beer that is average in alcohol content.

Mash Temperature and Thermometers

The temperature of the mash contributes to thickening the body of the beer more than any other factor. There is a very narrow window of temperature that changes the fermentability of the wort. For every degree Fahrenheit above 151 the wort ferments 1.5% less.[8] At 160, only 9 degrees higher, the enzymes that convert starch to sugar can be destroyed to the point that they will not complete the conversion of the grain into fermentable wort. Not only is even temperature of the mash critical for this reason, but accurate temperature is very important as well. In my kitchen there are three thermometers, and they all measure boiling water as different temperatures. Water, at one atmosphere with no additions, boils at 212°F by definition. The three thermometers measured 213, 216 and 217 degrees! The thermometer from the grocery store was off by 5 degrees.

While this may not seem like a tragedy, it can be. According to Greg Noonan, author of "New Brewing Lager Beer," for a light beer the target saccharification temperature is 150°F, and for a malty, full bodied beer the target is 157°F. That's only a seven degree difference from light to full body.

In three mashes I targeted 158 degrees to enhance the body of the beer. Because my thermometer was reading high by 5 degrees the actual mash temperature was 153 degrees which is the target for typical ales. Needless to say it didn't come out to the thick malty style that was intended. Had the target been toward the low end, to create a highly fermentable wort, it is likely that there would not have been complete conversion. This would leave starch in the wort as well as create a low initial gravity and a low efficiency.

So what can you do about it?

---

[8] See " Measured Mash Temperature Effects" Section

1) Buy a better thermometer. Pen style thermometers seem to be more accurate than the oven style out of the box in my experience. The Tru Temp that my wife purchased is very accurate. If you want to be certain then purchasing a thermometer that has calibration data will let you know how far it is off. Many brewers swear by the Thermapen. It is pricy, so I haven't tried one, but this does have calibration data: +/- 2°F. Also, something with a Type K or J thermal couple is probably a good place to start.

2) Compensate for the thermometer's inaccuracy. The two easiest known temperatures to measure are boiling and freezing. Mash temperatures are closer to boiling. So, if you boil water, and live near sea level, you should measure 212°F (100°C). If you know how many degrees off your thermometer is you can correct the value that is read.

3) Calibrate your thermometer. This is most easily done with an oven style thermometer that can be unplugged. If you're an electrical engineering geek like me, this is a fun little project. Measure the resistance of the thermal probe at freezing and boiling. Record what the thermometer reads under these conditions. Calculate the required trimming potentiometers to add in parallel with the thermal probe and in series. Trim the large parallel resistor to make ice water read 32°F, then trim the small series resistor to make boiling water read 212°F. Check again at 32°F and repeat if needed. It looks like for my oven thermometer a 10 Meg trim pot in parallel and a 1 k trim pot in series will do the trick.

Calibrating your Thermometer

Every measurement device has some inherent error.

An oven thermometer can easily be off by five degrees! This doesn't mean that it is "broken" or "bad," simply that it is not accurate. Of course a display that is closer to actual temperature will not only make use of the thermometer easier, but it will give you confidence that it is accurate at temperatures that are harder to measure for certain.

Measuring the error

Two temperatures that are easy to recreate accurately are 32°F (0°C) and 212°F (100°C) To measure the error, bring water to a rolling boil and measure the temperature. One particular thermometer measured 210.0°F, or two degrees low. Liquid water can only get to 212°F before it vaporizes, making it impossible to heat the water too much. Solid ice, however, can be colder than 32°F. If the thermometer were placed in a freezer in a glass of water the temperature measure could be below 32°F. Instead, the thermometer can be placed in a glass of ice water. More ice will get the water down to freezing faster, as will putting the glass in the freezer, but you'll want to take the measurement before the water freezes.

A third temperature can be recorded that is between these two. It's not necessary, but will give you a little more data to work with and may give you an idea of the accuracy of the interpolated temperatures between. If you don't have a fever, then your body temperature, under the tongue, should be 98.2+/-0.7°F mid-morning.[9]

The simplest approach is to use the deviation at boiling for mash temperatures and use the deviation at freezing for fermentation temperatures.

---

[9] http://en.wikipedia.org/wiki/Human_body_temperature

Crunching the numbers.

But if you are like me and want to really geek this out, keep reading, and don't worry, you only have to do this once. The recorded deviation at boiling and at freezing can be combined to create a reasonable guess of the deviation of temperatures measured in between. For this we need to derive a best line fit for the boiling and freezing points, and then we can use this to check the third temperature point that was measured to get a sense of accuracy.

We know two situations and have two unknowns, so this is solvable. I'll spare you the bore of watching me hash out the algebra, and skip to the equations:

BCF = Boiling Correction Factor
FCF = Freezing Correction Factor
MBT = Measured Boiling Temperature
MFT = Measured Freezing Temperature
CF = Correction Factor
MT = Measured Temperature
m = slope of fit
b = y intercept of fit.

CF = m * MT + b

m = (BCF-FCF)/(MBT-MFT)
b = CF - m * MT

For example, my thermometer measured 210°F at boiling, and 33°F degrees at freezing. My body temperature measured 98.5°F.

BCF = Boiling Correction Factor = 2
FCF = Freezing Correction Factor = -1

m = (2-(-1))/(210-33) = 0.0169

b = 2 - 0.0169*210=-1.549

We can then use this to find the correction factor at 98.5°F

0.0169*98.5-1.549=0.1157

The actual temperature would be 98.5+0.1157=98.6157 -- not bad.

To avoid this complicated math every time there thermometer is used an abbreviated table with correction factors for common temperatures can be affixed to the thermometer.

BREWING ENGINEERING

Mash Temperature Theory

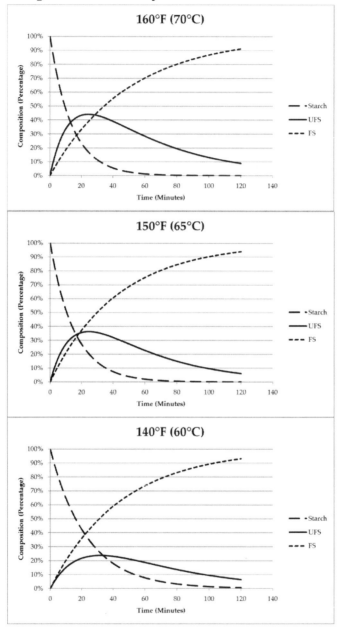

Starch, Unfermentable and Fermentable Sugars

Most likely you've heard about the enzymes in wort before, but I'm going to try to explain from a somewhat different perspective that hopefully sheds some new light onto the subject for you. It did for me.

There are two enzymes that primarily work during the saccharification of the wort. These are Alpha-amylase and Beta-amylase. The Betas work slowly, produce easily fermentable sugars, and work best at lower temperatures. The Alphas work faster, produce hard to ferment sugars, and work best at higher temperatures. You can see from this that selecting a mash temperature and time will change how much work each of these enzymes can perform.

At low mash temperatures, as seen in the 140°F chart above, the Alpha enzymes are barely active. The Beta enzymes, however, are highly active, making up for the difference causing a 140°F saccharification mash rest to produce about as many fermentables as a 150°F rest. At higher temperatures the Beta enzymes are not as active, but the Alpha enzymes are very active. This produces a mash that has 100% conversion, but has less fermentables remaining.

Measured Mash Temperature Effects

It's pretty well documented that a higher temperature mash will yield less fermentable wort, but to fine-tune my recipes correlating mash temperature to attenuation is important.

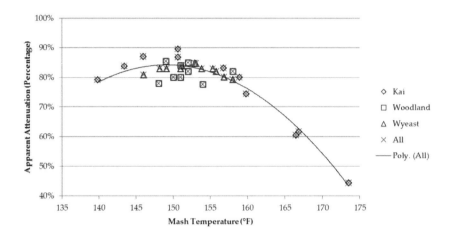

The above graph is a culmination of data from three sources. Two are independently run tests isolating mash temperature holding all other variables as contestant as possible. The diamonds are data from Kai Troester.[10] The triangles are from a study done by Wyeast. In addition to these two controlled studies there are several batches of beer that I have brewed to add a "real world" comparison. These are the squares.

You can see that the three all have a very similar shape and are roughly the same, however it's how they deviate that I find interesting.

Kai's data shows a maximum attenuation at about 148°F while the Wyeast data shows maximum attenuation at 153°F. There aren't really enough data points in my data to draw a conclusion here, but the peak seems to be about where Kai's is.

---

[10] http://braukaiser.com/wiki/index.php?title=Effects_of_mash_parameters_on_fermentability_and_efficiency_in_single_infusion_mashing

The diastatic power of the malt, mash time, and yeast strain all play into the equation as well, and likely account for these deviations.

It can be seen from this data that in the typical mash range of 145°F-160°F the attenuation can be varied by 10%. This is the largest contributing factor that I have seen. Yeast strain can vary attenuation about 4%. The use of crystal malt in a typical 10% addition changes attenuation about 3%. Mash thickness and fermentation temperature do not seem to affect final attenuation by a noticeable amount.

If you use a spread sheet for calculating attenuation then it's easy enough to add the full equation, but a basic rule of thumb will work for most cases. There are so many factors involved here that this rule of thumb might get you closer than the full equation.

Rule of thumb:

1% less attenuation for every degree above 151 (starting with the yeast's average attenuation)

The equation :

$$-0.000663 \times temperature^2 + 0.1964 \times temperature - 13.692$$

Predicting Brewhouse Efficiency

Efficiency can seem like a very complicated subject, but if you boil it down to the basics, it's really quite simple.

Conversion efficiency -- enzymes converting starch to sugar -- is near 100%. This is the case when mash time is one hour or longer. Spillage is minimal. This leaves lautering efficiency as the primary driver for brewhouse efficiency.

The only sugar that doesn't make it to the boil kettle is tied up in the grain, or lost in the dead space of the tun. The mash must be stirred well to disperse the sugars equally throughout the volume of water in the mash tun. This makes the efficiency of a no sparge mash a simple ratio. It is the amount of wort lautered from the mash, divided by the total volume of water that was added to the mash. Pre-boil volume, or wort removed from the mash, is the total water added, less the amount absorbed by the grain. Typically 0.15 gallons of water are absorbed per pound of grain. This can all be written in an equation:

E = Efficiency as a decimal value (e.g. a result of 0.10 would mean 10%)
W = Water used in mash, measured in gallons
G = Grain weight, in pounds

$$E = \frac{W - 0.15G}{W}$$

In practice, using this equation, or the table shown here, has been more accurate than measuring the gravity of the wort. Gravity readings taken on brew day are notoriously difficult. The temperature of the wort can skew readings. Even after compensating for temperature, there is plenty of room for error. How well the wort is mixed can have an enormous effect on efficiency. This is one of the things I didn't believe until I saw it firsthand. Also, conversion may not actually be complete, even after mash out. In my experience things don't really seem to stabilize until the wort is cooled and in the fermentor, and at that point it's more difficult to make adjustments to the brew.

Grain Weight		Water in Mash (Gallons)									
lbs	kg	3	4	5	6	7	8	9	10	11	12	13
3	1.4	85%	89%	91%	93%	94%	94%	95%	96%	96%	96%	97%
4	1.8	80%	85%	88%	90%	91%	93%	93%	94%	95%	95%	95%
5	2.3	75%	81%	85%	88%	89%	91%	92%	93%	93%	94%	94%
6	2.7	70%	78%	82%	85%	87%	89%	90%	91%	92%	93%	93%
7	3.2	65%	74%	79%	83%	85%	87%	88%	90%	90%	91%	92%
8	3.6	60%	70%	76%	80%	83%	85%	87%	88%	89%	90%	91%
9	4.1	55%	66%	73%	78%	81%	83%	85%	87%	88%	89%	90%
10	4.5	50%	63%	70%	75%	79%	81%	83%	85%	86%	88%	88%
11	5.0	45%	59%	67%	73%	76%	79%	82%	84%	85%	86%	87%
12	5.4	40%	55%	64%	70%	74%	78%	80%	82%	84%	85%	86%
13	5.9	35%	51%	61%	68%	72%	76%	78%	81%	82%	84%	85%
14	6.4	30%	48%	58%	65%	70%	74%	77%	79%	81%	83%	84%
15	6.8	25%	44%	55%	63%	68%	72%	75%	78%	80%	81%	83%
16	7.3	20%	40%	52%	60%	66%	70%	73%	76%	78%	80%	82%
17	7.7	15%	36%	49%	58%	64%	68%	72%	75%	77%	79%	80%
18	8.2	10%	33%	46%	55%	61%	66%	70%	73%	75%	78%	79%
19	8.6	5%	29%	43%	53%	59%	64%	68%	72%	74%	76%	78%
20	9.1		25%	40%	50%	57%	63%	67%	70%	73%	75%	77%
21	9.5		21%	37%	48%	55%	61%	65%	69%	71%	74%	76%
22	10.0		18%	34%	45%	53%	59%	63%	67%	70%	73%	75%
23	10.4		14%	31%	43%	51%	57%	62%	66%	69%	71%	73%
		12	16	20	24	28	32	36	40	44	48	52
		Water in Mash (Quarts or Liters)										

Table 10 - Efficiency From Water Volume and Grain Weight

You can get a few more points out by sparging. One sparge will add 6% for a 1.020 OG beer, 7% for a 1.030 OG beer and 9% for a 1.060 OG beer.

Why this is all you need to know

This equation is for no-sparge efficiency. As you increase the number of sparges, the efficiency will go up, but not by as much as some might hope. For each sparge, the increase in efficiency can be maximized by making the volume of the sparge equal to that of the run-off. So for one sparge, if you want 6 gallons of wort pre-boil, your first runnings should be 3 gallons and your sparge should also be 3 gallons. Each time you sparge you are diluting the wort that is tied up in the grains and getting the same efficiency on that sugar as you did on the first runnings.

More Grain Isn't Always More Sugar

It's common knowledge that if you want more fermentables, then you should add more grain. It makes simple sense, but every brewing system has its limitation. Regardless of the system you use, there is a point when adding more grain does not add more fermentables. In fact, eventually adding more grain will result in less fermentables.

You read that right: More grain can equal less fermentables!

The limit comes when the water absorbed by the grain approaches the total water added to the mash. Grain absorbs about 0.8 quarts of water per pound. So, as your mash out thickness approaches this, more and more of the converted sugars are trapped with the water absorbed by the grain.

For example we can compare 9 pounds of grain to 7 pounds. It would be easy to assume that using 9 pounds of grain would produce more sugar in the final wort, but this is not always the case.

For example, if your mash tun holds 4 gallons, 9 pounds of grains will displace 0.72 gallons. The maximum water that can be added is 3.03 gallons leaving 0.25 gallon head space for stirring. Grain absorption will be 1.8 gallons, making the first runnings 1.23 gallons of wort. This is 40% of the water that was added meaning that only 40% of the extracted sugars are in these runnings. 40% * 34 GU/lb * 9 lbs = 122.4 gravity units of sugar.

If only 7 pounds of grain were used, they would displace 0.56 gallons. Leaving the same space for stirring as in the previous example, the limit to the water added is slightly more at 3.19 gallons. Grain absorption is a little less at 1.4 gallons. This makes the first runnings a fair amount higher at 1.79 gallons. The runnings are 56% of the total water. 56% * 34 * 7 = 133.28 gravity units which is about 11 points higher!

BREWING ENGINEERING

With no-sparge, it is important not to overload your mash tun. The table below shows the extracted sugar in gravity points. Notice that the total sugar increases with the weight of grain and then decreases when the tun begins to be overloaded. The boldface numbers are the maximum sugar extracted for every gallon of water

grain weight		\multicolumn{11}{c}{Water in the mash (Gallons)}										
lbs	kg	1	2	3	4	5	6	7	8	9	10	11
2	0.9	48	58	61	63	64	65	65	65	66	66	66
4	1.8	**54**	95	109	116	120	122	124	126	127	128	129
6	2.7	20	**112**	143	158	167	173	178	181	184	186	187
8	3.6		109	163	190	207	218	225	231	236	239	242
10	4.5		85	**170**	213	238	255	267	276	283	289	294
12	5.4		41	163	224	261	286	303	316	326	335	341
14	6.4			143	**226**	276	309	333	351	365	376	385
16	7.3			109	218	**283**	326	357	381	399	413	425
18	8.2			61	199	282	337	376	405	428	447	462
20	9.1				170	272	**340**	389	425	453	476	495
22	10.0				131	254	337	395	439	474	501	524
24	10.9				82	228	326	**396**	449	490	522	549
26	11.8				22	194	309	391	**453**	501	539	571
28	12.7					152	286	381	452	508	552	589
30	13.6					102	255	364	446	**510**	561	603
32	14.5					44	218	342	435	508	**566**	613
34	15.4						173	314	419	501	**566**	620
36	16.3						122	280	398	490	563	**623**
38	17.2						65	240	371	474	556	**623**
40	18.1							194	340	453	544	618
42	19.1							143	303	428	528	610
		4	8	12	16	20	24	28	32	36	40	44

Water in the mash (Quarts or Liters)

Table 11 - Maximum Extracted Fermentables

For any volume of water that can be added to the mash, there is a maximum amount of grain that can be used effectively. As can be seen in the table, this occurs at 3.3 pound of grain per gallon of water or 0.4 kg per liter.

Maximizing Your Mash

The volume of wort removed from the mash tun is limited by both the size of the tun and the amount of grain. Each pound of grain displaces 0.08 gallons in the tun. In addition to the displacement of the grain, there is also absorption of water by the grains. These two factors combined limit the amount of water that you will get in your runnings. From the table below you can see that if you had a 44 quart, or 11 gallon, mash tun with 14 pounds of grain the maximum pre-boil volume of wort that you could obtain would be 7.5 gallons. It would likely be difficult to manage a mash tun full to the brim, so in this example six and a half gallons may be more reasonable.

Grain Weight		\multicolumn{11}{c}{Lauter Tun Volume (gallons)}											
lbs	kg	4	5	6	7	8	9	10	11	12	13	14
2	0.9	3.5	4.5	5.5	6.5	7.5	8.5	9.5	10.5	11.5	12.5	13.5
4	1.8	3.1	4.1	5.1	6.1	7.1	8.1	9.1	10.1	11.1	12.1	13.1
6	2.7	2.6	3.6	4.6	5.6	6.6	7.6	8.6	9.6	10.6	11.6	12.6
8	3.6	2.2	3.2	4.2	5.2	6.2	7.2	8.2	9.2	10.2	11.2	12.2
10	4.5	1.7	2.7	3.7	4.7	5.7	6.7	7.7	8.7	9.7	10.7	11.7
12	5.4	1.2	2.2	3.2	4.2	5.2	6.2	7.2	8.2	9.2	10.2	11.2
14	6.4	0.8	1.8	2.8	3.8	4.8	5.8	6.8	7.8	8.8	9.8	10.8
16	7.3	0.3	1.3	2.3	3.3	4.3	5.3	6.3	7.3	8.3	9.3	10.3
18	8.2		0.9	1.9	2.9	3.9	4.9	5.9	6.9	7.9	8.9	9.9
20	9.1		0.4	1.4	2.4	3.4	4.4	5.4	6.4	7.4	8.4	9.4
22	10.0			0.9	1.9	2.9	3.9	4.9	5.9	6.9	7.9	8.9
24	10.9			0.5	1.5	2.5	3.5	4.5	5.5	6.5	7.5	8.5
26	11.8				1.0	2.0	3.0	4.0	5.0	6.0	7.0	8.0
28	12.7				0.6	1.6	2.6	3.6	4.6	5.6	6.6	7.6
30	13.6					1.1	2.1	3.1	4.1	5.1	6.1	7.1
32	14.5					0.6	1.6	2.6	3.6	4.6	5.6	6.6
34	15.4					0.2	1.2	2.2	3.2	4.2	5.2	6.2
36	16.3						0.7	1.7	2.7	3.7	4.7	5.7
38	17.2						0.3	1.3	2.3	3.3	4.3	5.3
40	18.1							0.8	1.8	2.8	3.8	4.8
42	19.1							0.3	1.3	2.3	3.3	4.3
		16	20	24	28	32	36	40	44	48	52	56

Lauter Tun Volume (quarts or liters)

Table 12 - Runnings From Tun Size and Grain Weight

But what if you want more?

If you are making a high gravity beer, chances are that you aren't getting as much wort out of the tun on your first running to make your pre-boil volume. That's where sparging comes into play. The easiest way to do it would be to add the amount of water needed to make up the pre-boil volume to the mash tun, give it a good stir and drain again. However, maximum efficiency can be achieved by keeping each portion of the runnings the same.

Because each volume of wort removed from the tun should be the same simply divide your pre-boil volume by the number of times you will drain the tun. That will be the amount of wort you need for each collection. For example, my indoor brew system is a 4 gallon BIAB. My target pre-boil volume is 3.5 gallons to leave a little room on top. If I drain the grains twice that means I need to collect 1.75 gallons each time. From the table we can see that is possible with up to 9 pounds of grain in my 4 gallon pot. If I drained 3 times it would be up to almost 12 pounds.

With a BIAB system, one simple way to sparge is by using a bucket. After the grains have been pulled out and drained they can be moved to a bucket. Measuring the volume of wort in the kettle will determine how much more water should be added to the grains for the sparge. The temperature of the sparge is not nearly as critical as the mash temperature, so hot tap water normally works well for this.

Wort Sugars

For yeast to ferment multi chain sugars, it must first be broken down into a single chain by an enzyme. Whether a yeast can produce these required enzymes is dependent on its genetics. So, if a typical wort is split into several different fermentations, and each is inoculated with different yeasts, each one should fall into a bin representing its enzyme capability.

Sugar Name	Chain Length	Fermentation Order	Wort Composition
Glucose	1	1	8%
Fructose	1	2	2%
Maltose	2	3	45%
Matotriose	3	4	14%
Dextirins	4 or more	Un-fermentable	25%

Table 13 - Wort Sugar Composition

See footnotes [11] [12]

To find out more precisely the sugar content of the dried malt extract you are using, search for the data sheet provided by the maltster. For example, one data sheet from Briess shows a similar breakdown to the one shown in the table above, although the fructose seems to be replaced by more glucose. This indicates that CBW® Golden Light DME is composed of 13% Glucose, 48% Maltose, 14% Maltotriose, and 19% Higher Saccharides.[13]

---

[11] http://www.brewmorebeer.com/brewing-sugars/
[12] http://en.wikipedia.org/wiki/Malt_sugar
[13] http://www.brewingwithbriess.com/Assets/PDFs/Briess_PISB_CBW_GoldenLightDME.pdf

Given this wort composition, I would expect fermenting microorganisms to fall into four categories:

10% attenuation

only capable of fermenting simple sugars.

55% attenuation

capable of fermenting 2 chain sugars but not 3 chain.
Eg. Wine yeast

75% attenuation

capable of fermenting 3 chain and shorter sugars.
Eg. Brewing yeast

100% attenuation

capable of fermenting all sugars.
Eg. Hyper attenuating bacteria and possibility Brett (Brettanomyces)

Different strains of yeast can be used to ferment wort in order to control the sugar content. To test this, an experiment was run using dry malt extract to create five worts of different gravities. These were split into a total of 15 cultures and were allowed to ferment to completion with three different yeasts. The brewer's yeast attenuated the wort 62%, while the wine yeast attenuated the wort 40%. From this it seems that the dry malt extract contains 40% one and two chain sugars, and 22% three chain sugars; the remaining 38% is four or larger chain sugars.

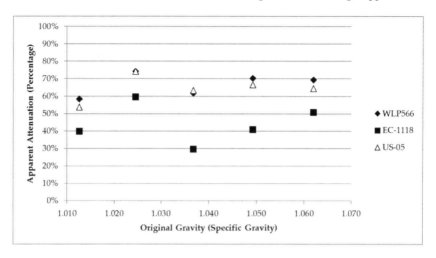

There are other factors that affect attenuation. Most notable is available oxygen. For more accurate and consistent results of this test you can use a stir plate or agitation table rather than culture tubes.

For recipe formulation this fast ferment test can be used to assess brewing ingredients and their impact on alcohol and final gravity in both cider and beer making.

Another useful, and perhaps more interesting, application of the ability of not all strains to ferment maltose would be to determine sugar composition produced by different mash temperatures. A sample could be pulled from the mash periodically and then quickly boiled to denature the enzymes. These samples could then be evaluated with a fermentation test to determine the sugar composition.

Hitting the Exact Original Gravity

Want to know a little secret? Brewers almost never hit their exact OG right out of the mash tun.

Original gravity is very important to the balance of a recipe. OG is essentially the sum of the alcohol and the residual sugar left in the beer. Both of these are commonly offset by hops. This is where the BU:GU ratio comes from.[14] If the OG is significantly low, the beer will have less alcohol and residual sugar which will cause the beer to taste more hop focused. The converse is also true. Recipes are balanced for a reason. Tipping the balance makes a different beer, which might be a fine beer, but the recipe was likely selected or designed with a goal in mind.

So if brewers don't hit their OG right away, how is it achieved?

There are several ways this can be done. This is my plan of attack:

1) If the OG is high, then add more water. If the OG is low then add DME.

2) If the first suggestion can't be done, then scale the hops.

Adding Top Off Water

Specific gravity is a ratio of dissolved solids (in this case, sugar) to water. It can, therefore, be simply scaled. I find it easiest to deal with the numbers if they are in gravity points. For example if I was trying to make a 4 gallon 1.050 wort, but my gravity was 1.060 because of unexpectedly high efficiency, I would need to add top off water.

4 gallons * 60 GP = x gallons * 50 GP =>

x gallons = 4 gallons * 60 / 50 => x = 4.8 gallons

---

[14] Ray Daniels, "Designing Great Beers"

4.8 gallons are needed, so 0.8 gallons of water would need to be added to achieve the desired 1.050 wort. This, of course, makes more beer so the hops will need to be scaled as well. (See Scaling the Hops below.)

Adding DME

On the other hand if the OG was low, the gravity will need to be adjusted up. Let's say the 4-gallon batch came out at 1.050 and the target was 1.060. The easiest way I have found to adjust the gravity up is to add dried malt extract. We simply need to figure out how many gravity units need to be added to the beer.

4 gallons * 50 GP = 200 GU (What we have)

4 gallons * 60 GU = 240 GU (What we want)

240 GU - 200 GU = 40 GU (how much we need to add)

40 GU / 46 ppg = 0.87 lbs

0.87 lbs * 16 = 14 oz.

Boiling down the wort

An alternative to adding DME would be to decrease the batch size by boiling the wort down. However, as with adding water, because the volume has changed, the hops will need to be scaled as well.

4 gallons * 50 GP = x gallons * 60 GP

4 gallons * 50 GP / 60 GP = x gallons

x gallons = 3.33 gallons would need to be the volume of the wort after boiling it down.

Brewing Engineering

Scaling the Hops

Changing the amount of hops will make a different beer, but it will still be a balanced recipe. In the event that the other methods cannot be used for some reason, this could be a way to keep the beer from slipping too far from the intended style. We'll use the same example, where our specific gravity coming into the boil is 1.050 but the target was 1.060. We'll scale the hops back to maintain the same BU:GU ratio. Let's assume that the 60 minute addition was 0.5 oz of hops, the 30 minute was 0.75 oz, and the 5 minute addition was 1 oz.

$$Scale\ factor = \frac{actual\ OG}{target\ OG}$$

$$\frac{50\ GP}{60\ GP} = 0.83$$

bittering hops: 0.5 oz * 0.83 = 0.415 oz
flavor hops: 0.75 * 0.83 = 0.625 oz
aroma hops: 1.0 * 0.83 = 0.83 oz

MORE THAN MALT

4

Fruit the Easy Way

While adding fruit directly to the secondary fermenter may be an easy way to get it into the fermenter, it makes it difficult to separate it from the beer later. The times that I have added fruit this way, my bottles have always ended up with some pieces of fruit in them. While the fruit chunks are not any sort of hindrance to the flavor, they are certainly not appealing to have floating around in a glass of beer.

Frozen fruit is an economical choice, and has additional benefits over fresh fruit as well. The freezing process breaks down the cell walls, allowing more of the fruit flavors into the beer. In addition, contamination is less of an issue because cold temperatures inhibit bacteriological growth. Although some people add frozen fruit directly to the fermenter I wouldn't recommend it for two reasons. First, the cellulose of the fruit will release the juice into the beer, but it will be replaced by beer in the process. I have had as much as 33% trub loss because of this issue. The second reason is to avoid floaters.

The best way I have found to add fruit to beer is to pasteurize the fruit. Not only does this process remove almost all of the cellulose, but also kills most any bacteria that may have somehow made it to this point.

1) Add the frozen fruit to a sauce pan.
2) Add 1 pint of water for every pound of fruit.
3) Apply medium high heat to the sauce pan stirring constantly and mashing. (A potato masher works great for this!)
4) Once the temperature has reached 160°F, turn off the heat.
5) Pour this mixture through a sanitized mesh strainer and into a second sanitized container and affix the lid.
6) Allow this to cool for about 20 minutes and then move it to the refrigerator.
7) After several hours in the refrigerator, pour the contents into the fermentor.

If the hot fruit is added directly to the fermentor there would be a significant increase in temperature. A half-gallon of 160°F water added to beer fermenting at 65°F would raise the temperature to 73°F. This would be a shock to the yeast which would trigger temperature excretion[15] as the yeast cells adapt to the new environment.

This is more work on the front end, but it will save work on the back end trying to separate beer from fruit. And, more importantly, it will yield a few more bottles of beer!

---

[15] George Fix "Principles of Brewing Science" Shock Excretion pg 90 2nd edition

How Much Hops

Beers, as described in the Beer Judge Certificate Program (BJCP) style book, range from zero to a puckering 120 International Bittering Units (IBUs). By evaluation the data provided by the BJCP (standard deviation, and mean average) these can be broken down into categories:

Lightly Hopped: 0-15 IBUs
Moderately low hops: 15-25 IBUs
Average: 25-35 IBUs
Moderately high hops: 35-45 IBUs
Highly hopped: 45 IBUs and up.

My preference is for the moderately low range, and my wife prefers lightly hopped beer. It's a fairly simple process to find the range you like and calculate how much hops to use to achieve that same bitterness level. This can easily be done with various calculators found on the Internet.

However, there is more to hops than just the bitterness level.

Over time as a brewer, you will get a feel for the amount of bittering, flavor and aroma hops preferred in a beer, but without that experience there is no way to really qualify or quantify what a beer may taste like in a given recipe. In "Designing Great Beers," Ray Daniels gives a wonderful explanation of the wide variety of factors that play into flavor and aroma, compared to the relatively simple reactions that create bitterness.

One simple way to quantify the type of bitterness the hops may add is to break them into two groups: Bittering and Flavor. Hops added early in the boil, spending more than 25 minutes in the kettle, will add mostly bitterness. Hops added later in the boil, spending less than 25 minutes will add mostly flavor. The ratio of bitter IBUs to flavor IBUs can be a way to compare recipes.

bitter:flavor
4:1 - mostly bitter.
2:1 - balanced.
1:1 - mostly flavor.

It's also important to develop a sense of flavor for the various hops. My personal favorite hop is Challenger. It works wonderfully as both a flavor and bittering hop. Taste is very subjective, but if this is kept in mind, hop schedules will be better evaluated leading to better formulated recipes.

For more details on this subject this web page is a wonderful resource:

http://www.babblehomebrewers.com/attachments/article/68/hopusage.pdf

Growing Yeast

5

Visualizing Growth

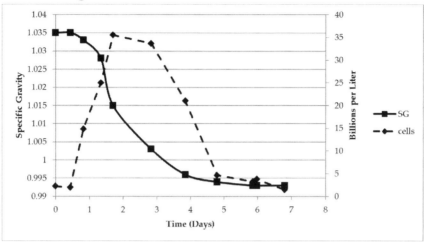

The graph above shows the suspended cell density and specific gravity of a normal fermentation. The fermentation was one gallon of 1.036 wort pitched with 7 billion cells per liter. The wort was 1 gallon composed of primarily sucrose and ½ tsp of yeast nutrient. The yeast pitched was 25 billion cells of EC-1118 Champagne Yeast.

Yeast Growth as a function of available sugar.

Note the hyper attenuation that resulted from this combination. Not only is champagne yeast known for its ability to fully ferment sugars, but when provided with yeast nutrients and a wort of simple sugars, it has no problem converting all of the sugar to alcohol. Because alcohol is lighter than water, the resulting solution is lower than 1.000 specific gravity.

Notice that the highest rate of cell production directly correlates to highest concentration of sugar. This is seen at the second through fifth data points. As the available sugar decreases, the cells fall out of suspension faster than they are produced (data points 6 through 8). When fermentation is nearly complete, the cells maintain a concentration of about 3 billion cells per liter.

The peak cell density suspended in the wort was 35 billion cells per liter, or about five times the initial cell count.

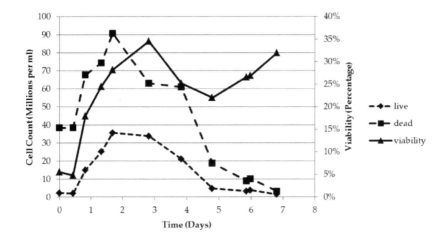

The yeast that was pitched was approximately 5% viable, as can be observed on the first point of the viability graph. After approximately half a day the yeast began dividing as can be seen by the increase in both viability and viable cell density.

The number of non-viable cells in suspension tracks that of the viable population. I suspect that this is caused by simple mechanical stimulation. As the living yeast cells produce CO_2, a small bubble forms on the cell causing it to float. On its way to the surface, it bumps into dead yeast cells and carries them upward.

The peak viability is reached at the same time the food source has been exhausted. As the fermentation completes, the viability percentage also increases, but the cell count in suspension is about a tenth of the peak value. So, although the non-viable cells do flocculate faster after fermentation has completed, there are very few viable cells on top of the layer of dead cells.

Starter Calculators

Quite a bit of work, especially recently it seems, has gone into ensuring that the correct amount of yeast is pitched to start fermentation. As a result, there are several calculators available that calculate how to build up the number of yeast cells required for optimal pitching rates. Considering the amount of work put into these calculators, the quality of their data sources, and the scientific reasoning that has gone into their development, one might expect them to be accurate.

The recommendations from three popular calculators are summarized here:

Estimator	Original Gravity	Billions of Cells	Size (liters)
Mr. Malty	1.050	175	1.40
Mr. Malty	1.080	273	3.90
Yeast Calc	1.050	175	1.37
Yeast Calc	1.080	274	3.64
Wyeast	1.050	114	0.34
Wyeast	1.080	227	1.61

Table 14 - Stater Calulator Summary

Mr. Malty and Yeast Calc are very close to each other in estimations, which we would expect, as both of them use the same set of data to create their calculations. The Wyeast calculator recommends approximately the same number of cells, but the starter size is much smaller. This leaves two questions:

1) Which one is right?
2) Does it matter?

In the remainder of this chapter these questions will be addressed.

Viability vs. Vitality

Viability - the percentage of live cells in the total cell population.

Vitality - a measure of yeast health.

Viability is a quantitative measurement, meaning it is something that can be counted. With methylene blue staining viability can be assessed fairly quickly. This makes it a very accessible way to measure yeast health.

However, it's important to remember that there is much more to yeast health than viability.

Vitality itself cannot be counted making it a qualitative measurement. However, there are aspects of vitality than can be measured and used to assess the overall health of the yeast.

Glycogen levels - Stored energy used in production of sterols. These are vital to cell membrane permeability. Glycogen is similar to starch and can be measured with an iodine test.

Lag phase length - Healthy yeast will have a very short lag phase because their cell membranes are already prepared for fermentation.

Ratio of mother cells to daughter cells – A skilled technician can approximate this using a microscope.

Fermentation products - These can vary fairly widely and are difficult to differentiate.

It's possible to have yeast that has high viability yet a low vitality (Plenty of living cells, but mostly of poor health). An example of this would be slurry that has been stored in the refrigerator for more than a week. Although there may be plenty of cells available, the wort will need to be aerated well to provide the oxygen necessary to build sterols, and the lag phase will be extended while these sterols are produced.

Measured Cell Growth

There are a few calculators available to estimate the number of cells grown simply based on the number of cells that you are starting with, and the volume of wort that is fermenting.

But how accurate are these calculators?

When I added a microscope to my brewing equipment I was surprised to learn how far off online calculators can be.

There are some factors that remain the same from one setup to the other, and some that change. Most people propagate cells at ambient air conditions, so in these experiments temperature will not be used as a variable. Also, not everyone has access to a stir plate, so no stir plate will be used in these experiments.

One of the things that people often seem to change without thinking about the consequences is the specific gravity of the wort. Because most starter calculators seem to be based on data collected with a 1.036 specific gravity wort, a range around this gravity will be used.

The next factor is the ratio of cells pitched to wort volume, also known as the inoculation rate. The yeast, being very small, have no idea how big the starter is. It could be 10 ml or 5 gallons, and they would function the same. What may effect yeast growth is how crowded the vessel is. This is essentially the inoculation rate. If there are 1 billion cells in a 10 ml wort, or 10 billion cells in a 100 ml wort the growth will be the same.

Another factor that is not considered by the calculators is the strain of yeast. By testing some common strains and some not so common ones a better idea of how each strain may perform can be evaluated.

A total of 50 tests were run.

A 5x5 matrix of specific gravity vs. inoculation rate. The Gravities were 1.010, 1.025, 1.036, 1.045, and 1.100 (2, 6, 9, 12, and 24 degrees Plato). The inoculation rates were 5, 55, 65, 75 and 200 million cells per ml.

Dr. Chris White at White Labs conducted similar experiments for the Mr. Malty calculator. This data was taken and also used for Yeast Calc. These calculators both show that an inoculation rate of 65 million cells per milliliter should exhibit the most growth per volume of wort. These tests will all use Safale S-04 as it is very widely used and commonly available.

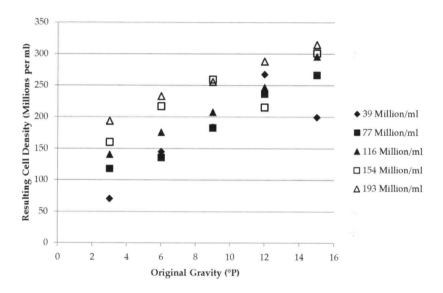

Cell growth followed fairly close to the "old school" rule of 10 billion new cells per liter per degree Plato. This is also roughly equivalent to 1 billion new cells per gram of DME. There was some correlation between inoculation rate and cell growth which may be an indication of crowding. The number of cells produced was roughly twice that predicted by the popular calculators.

Cell death can also play into the number of viable cells in the starter. If the yeast are allowed to sit in alcohol for a period of time after growth they will begin to die. See ABV effects on alcohol.

Yeast Growth Rate

Three factors that drive the rate of cell propagation and fermentation are the strain of yeast, temperature, and vitality. If these factors are constant then the rate of reaction can be greatly simplified.

During fermentation yeast is suspended in a sugar solution. Yeast begins to metabolize sugar on contact. Fermentation can be thought of as a biological reaction. It follows then that the rate of this reaction is driven by the concentration of both yeast and sugar. For typical fermentation if the cell count is doubled, then the chance that the yeast will come into contact with sugar is also doubled.

Some sugars are more difficult for yeast to metabolize than others. Single chain sugars such as glucose and fructose are rapidly converted, while two chain sugars such as maltose can take much longer. The rate at which these sugars are metabolized can be observed experimentally.

Yeast stays suspended in the wort because it creates CO_2 gas. These bubbles cling to the cell propelling it upwards until the bubble is either large enough to disengage from the yeast or the yeast reaches the surface where the gas is released. The yeast then falls toward the bottom of the fermentor until it has produced enough CO_2 to gain buoyancy. Once the sugar is depleted there will be no CO_2 production and the yeast will slowly sink.

Using this growth theory a model was constructed and compared to real fermentation for validation. The fermentation was a 1.056 (14°P) wort of Amber Briess DME pitched with Fermentis WB-33. The cell count was verified with a hemacytometer using Methylene Blue for viability staining after rehydration. Temperature was maintained at 70°F (20°C). In the graph below the squares and diamonds represent measured values while the lines represent the prediction model. It can be seen that the model fits this fermentation well.

BREWING ENGINEERING

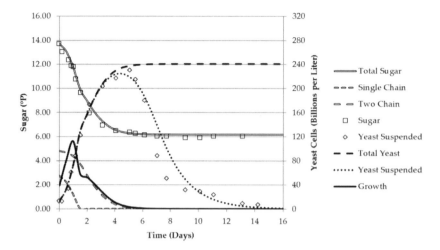

The "Total Sugar" predicted tracks the "Sugar" measured in the fermentation very closely.

Two portions of the growth phase can be identified. First, the rapid growth that takes only half a day, and then the slower growth that takes several days to complete. Without modeling these two growth rates the fit would not be nearly as close.

This model can be applied to beer fermentation. Using a standard ale pitch rate of 750 million cells per liter degree Plato some observations can be made.

- Cell population doubles in the first 6 hours as simple sugars are consumed.
- Doubling the pitch rate may reduce turnover rate by one day.
- Higher gravity beers, when adequately pitched, will actually ferment faster than low gravity beers.

The rapid increase in cell population directly after inoculation can be used to make larger beers without a starter. See "Do the Two Step" in the next chapter for more information.

Making Great Beer Through Applied Science

In addition to beer fermentation this model can be applied to starters. Results will vary depending on a number of factors, but this should provide a baseline for estimating cell count. Typical starters use 100 billion cells in 1.036 (9°P) wort. The number of cells based on time are summarized below.

		\multicolumn{6}{c}{Time (Hours)}					
cups	liters	12	18	24	30	36	42	48
1	0.2	156	160	161	162	162	162	162
2	0.5	159	168	174	178	181	183	184
3	0.7	163	174	183	190	195	200	203
4	0.9	177	190	201	211	219	226	231
5	1.2	185	204	218	230	240	249	256
6	1.4	191	218	234	248	260	271	280
7	1.7	195	229	249	265	280	292	304
8	1.9	198	238	264	282	299	313	326
9	2.1	200	245	277	299	317	333	348
10	2.4	202	252	288	314	335	353	370
11	2.6	204	257	299	329	352	372	391
12	2.8	205	262	309	343	369	391	411

Table 15 - Start Cell Count Over Time

Anaerobic and Aerobic Respiration

The primary goal of fermenting beer is the production of alcohol; while the goal of cell propagation is increasing the yeast biomass. What separates these two is oxygen. In oxygen-depleted environments yeast will be forced into anaerobic respiration which produces alcohol. On the other hand, with adequate oxygen yeast will favor aerobic respiration which is conducive to cell propagation.

The two things that yeast need from the wort to make new cells is material (sugar) and energy. While both of these are available during both aerobic and anaerobic respiration there is much more energy during aerobic respiration. A stir plate provides constant oxygenation which makes it well suited for starters.

Oxygen is good for propagation, but how much is required, and how can it be used to maximize cell growth?

Aerobic yeast respiration is as follows:

$C_6H_{12}O_6 + 6\ O_2 \rightarrow 6\ CO_2 + 6\ H_2O + 31\ ATP$[16]

$180g\ C_6H_{12}O_6 + 192g\ O_2 \rightarrow 264g\ CO_2 + 108g\ H_2O + 31\ ATP$

Converting from moles to grams we can see that for every gram of fermentable extract 1.07 grams of oxygen are required for aerobic respiration. For a 10°P (1.040) wort that would be a whopping 107,000 ppm! With pure oxygen gas the saturation point of water is only 50 ppm. So in terms of aerobic respiration, there is no practical limit to the amount of oxygen that can be utilized. Oxygen, however, is toxic to yeast in high concentrations.

Because oxygen is always in short supply anaerobic respiration dominates the metabolic activities. For this reason, the anaerobic

---

[16] http://en.wikipedia.org/wiki/Cellular_respiration

reaction very closely resembles Balling's observation. When the reaction is converted to moles it can be seen that the "losses" that Balling describes are the sugar converting to other materials.

Anaerobic Respiration:

$C_6H_{12}O_6 \rightarrow 2\ CH_3CH_2OH + 2\ CO_2 + 2\ ATP$[17]

1.9553g $C_6H_{12}O_6 \rightarrow$ 1g CH_3CH_2OH + 0.9553g CO_2 + 2 units ATP

Balling Observation:

2.0665g $C_6H_{12}O_6 \rightarrow$ 1g CH_3CH_2OH + 0.9565g CO_2 + 0.11g biomass [18]

The amount of Carbon Dioxide is virtually identical and the difference in glucose mass is almost exactly the yeast biomass.

If the two things that yeast need are sugar and energy to reproduce then how much more yeast would be generated from aerobic respiration? For every 2 units of ATP 0.11 grams of yeast are generated. If there were 31 units of ATP then 1.705g of yeast could be generated. This would require 3.6603g of sugar and adequate oxygen. Anaerobic respiration produces 0.053g of yeast per gram of fermentable extract. At 20 billion cells (dry mass) per gram that's 1 billion cells generated per gram of extract. Aerobic respiration could produce 0.4658g of yeast per gram of fermentable extract. This makes 9.316 billion cells of yeast produced per gram of extract.

---

[17] http://en.wikipedia.org/wiki/Ethonal
[18] (3) Balling C. J. N. 1865. "Die Bierbrauerei" Verlag von Friedrich Temski, Prague, CHZ. As cited in:
MODELING OF ALCOHOL FERMENTATION IN BREWING – SOMEPRACTICAL APPROACHES, Ivan Parcunev, Vessela Naydenova, Georgi Kostov, Yanislav Yanakiev, Zhivka Popova, Maria Kaneva, Ivan Ignatov http://www.scs-europe.net/conf/ecms2012/ecms2012%20accepted%20papers/mct_ECMS_0032.pdf

With adequate oxygen the yeast propagation could be almost tenfold what is typical of fermenting beer!

Yeast biomass production with a stir plate can therefore be quite unpredictable. The level at which aerobic respiration dominates the reaction is directly driven by the amount of oxygen available to the yeast. Producing twice the number of cells on a stir plate compared to a still starter is common.

Extrapolated Data

Data was collected on a set of 42 side by side starters. The intent of this experiment was to compare the performance of yeast that was actively fermenting in a starter to that of yeast taken from refrigerated slurry. To compare the performance across a variety of typical applications both inoculation rate and original gravity were also varied.

The Daily Data

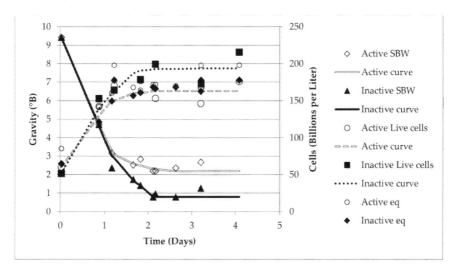

Two 15 ml test tubes were prepared. Both had a starting gravity of 9.42°P. One was inoculated with 58 million cells per milliliter taken from an active starter. The second was inoculated with 60 million cells per milliliter from slurry that had been in the refrigerator for one month. Measurements were taken approximately daily and the date and time were recorded with each measurement. These measurements included a refractometer measurement and a cell count with viability staining. The "active" description represents cells taken from the active starter, while the "inactive" represents cells taken from the refrigerator. SBW is the sugar by weight. This is the converted refractometer measurement using the equations derived earlier. The "eq" plots are a simple equation to show the relationship between consumed sugar and

BREWING ENGINEERING

cells produced. For the active culture this is 10 times the sugar consumed, and for the inactive it is 15 times the sugar consumed.

Initial Sugar (Percent sugar by weight)

	SO4A-2	SO4A-3	SO4A-4	SO4A-5	SO4I-1	SO4I-2	SO4I-3	SO4I-4
15	15.7	15.7	15.7	15.7	15.7	15.7	15.7	15.7
12	12.56	12.56	12.56	12.56	12.56	12.56	12.56	12.56
9	9.42	9.42	9.42	9.42	9.42	9.42	9.42	9.42
6	6.28	6.28	6.28	6.28	6.28	6.28	6.28	6.28
3	3.14	3.14	3.14	3.14	3.14	3.14	3.14	3.14

Initial Cells (Millions in a 10ml test tube)

	SO4A-2	SO4A-3	SO4A-4	SO4A-5	SO4I-1	SO4I-2	SO4I-3	SO4I-4
15	290	435	580	725	301	602	903	1204
12	290	435	580	725	301	602	903	1204
9	290	435	580	725	301	602	903	1204
6	290	435	580	725	301	602	903	1204
3	290	435	580	725	301	602	903	1204

Final Cells (Milllions in a 10ml test tube)

	SO4A-2	SO4A-3	SO4A-4	SO4A-5	SO4I-1	SO4I-2	SO4I-3	SO4I-4
15	1282	1452	1825	2017	2238	2160	2683	2815
12	1201	1410	1403	1532	1678	1841	2073	2176
9	1031	1282	1275	1230	1317	1336	1707	2006
6	792	926	1021	1249	1275	1340	1630	1961
3	575	751	816	1140	900	1045	1262	1505

Table 16 - Fermentation Matrix Results

The five by eight matrix of tubes was set up to compare the results of varying inoculation rates from 30 million cells per milliliter to 120 million cells per milliliter. The wort gravity was varied from 3°P to 15°P (1.012 to 1.060). In addition yeast was used from a starter at high krausen for one half and yeast that had been refrigerated for the other half.

To achieve the cell counts, each tube was counted and then varying volumes were added to the culture tubes. For the active yeast cells these ranged from 2 to 5 ml, and for the inactive culture they ranged from 1 to 4 ml. Adjustments to gravity measurements are needed because there is inherently residual sugar and alcohol in the slurry.

The array of gravities was achieved by diluting a 31.4°P wort and adding 1 to 5 ml. The remaining volume was filled with water to ensure each tube had 10 ml of total volume.

The tubes were allowed to ferment for two weeks and then refrigerated and allowed to settle for several days. The height of the yeast cake in each tube was measured. Six cell counts were done and these were correlated to the heights of the remainder of the tubes to produce the final cell counts. A linear fit based on the number of ml of slurry was used to equate milliliters of slurry to number of cells. For this fit, the r^2 value for the active culture was 0.9936 and the r^2 value for the inactive culture was 0.9850.

Observations of the daily collected data.

When looking at the daily collected data, the first thing that really jumped out was the correlation between sugar consumed and cell count. Sugar is known to be a limiting factor in cell propagation, but seeing how well the two correlated was surprising. This opens up a new way to look at yeast growth. Trying to determine cell growth solely from the input conditions is not using all of the data available. If in addition the final sugar by weight is used a much better approximation of cell count can be achieved.

Both of the test vials produced very close to 1.2 billion cells for every gram of extract consumed. Another way to look at that number is relative to volume. For each degree Plato the cell density increases by 1.2 million per ml.

Comparing the two

Both the refrigerated cells and the new cells started consuming sugar almost immediately. Both had reduced the sugar in the wort by half in the first day.

The daily data shows that the refrigerated slurry out-performed the cells removed from a starter. This was a second big surprise. Common brewing knowledge would indicate that cells that have been in the refrigerator for a month will be starved, and will not perform well, however quite the contrary was the case here.

This unusual performance may be linked to glycogen reserves. At the start of fermentation yeast will build glycogen reserves. During the growth phase these are significantly depleted during cell division. At the end of fermentation the yeast will rebuild these reserves as they prepare for dormancy.[19] The cells taken from the refrigerator were allowed to ferment to completion, and even after a month in the refrigerator still had significant glycogen reserves to support cell division.

It seems that it is better to allow a starter to run to completion than to use the cells at high krausen.

There are three big takeaways from this collection of data.

1) For reasonable inoculation rates and gravities cell growth is not a function of inoculation rate or volume. Cell growth is simply a direct function of sugar, and may be more accurately predicted by observing sugar consummation.

2) Yeast taken from storage at 40°F (5°C) can out-perform yeast taken from an active starter likely due to the glycogen content of the yeast.

---

[19] Fix, Principles of Brewing Science, p97 in the 2nd addition

3) Attenuation is a function of both inoculation rate and gravity. Higher inoculation rates and lower gravities lead to higher attenuation.

Growth

Cell growth observed over this wide range of inoculation rates and initial gravities was proportional to the amount of initial sugar present. In the case of the active culture of S-04, 7.2 billion cells were grown per liter per initial degree Plato. Considering that the average attenuation was 62% this indicates that 11.6 billion cells were grown per liter degree Plato of consumed sugar. This tracks very well with the daily observations. In a similar fashion the inactive starters grew 10.4 billion cells per liter per initial degree Plato with an average attenuation of 68% making 15.3 billion cells grown per liter degree Plato of sugar consumed. Both of these starters closely followed the equations derived from Balling's observation for starter cell growth. Based on these observations the equation can be adjusted to account for the higher cell growth that is likely linked to proper aeration at the onset of inoculation.

Cells Grown (Billions) = 14 * Volume of wort (Liters) * [Initial Gravity of wort (°P) - Final Gravity of wort (°P)]

Glycogen

The yeast taken from refrigeration out-performed the yeast taken from a starter in terms of total attenuation and cell growth. When fermentation is allowed to run to completion yeast cells will build up a glycogen reserve. It is possible that this extra glycogen allowed the refrigerated yeast to grow more cells given the same amount of sugar. Further testing would be required to confirm this, including measurement of the glycogen levels.

Attenuation

The percentage of sugar consumed seems to be a function of both the initial gravity and the inoculation rate. A linear fit for these two parameters shows an excellent r squared fit of 0.9139 and 0.9795. Although this shows wonderful correlation I am hesitant to say there is causation. There are other considerable factors, such as temperature, that are not accounted for here. These equations work very well for inoculation rates typically used for starters, but may not work to predict attenuation in beer.

°P	\multicolumn{8}{c}{Innoculation Rate (millions of cells per milliliter)}
	29	30.1	43.5	58	60.2	72.5	90.3	120.4
15	55%	57%	58%	57%	58%	53%	61%	62%
12	56%	55%	56%	61%	61%	61%	63%	63%
9	58%	58%	61%	65%	62%	64%	65%	67%
6	57%	61%	60%	69%	62%	70%	73%	73%
3	66%	59%	67%	73%	69%	81%	76%	83%

Actual Attenuation

°P	\multicolumn{8}{c}{Initial Cells (Millions in a 10ml test tube)}
	29	30.1	43.5	58	60.2	72.5	90.3	120.4
15	77%	80%	82%	80%	82%	75%	86%	87%
12	79%	77%	79%	86%	86%	86%	89%	89%
9	82%	82%	86%	92%	87%	90%	92%	94%
6	80%	86%	85%	97%	87%	99%	103%	103%
3	93%	83%	94%	103%	97%	114%	107%	117%

Apparent Attenuation

Table 17 - Actual Attenuation from Innoculation Rate and OG

The following equation can be derived from the linear fits:

A = Actual Attenuation (as a decimal. i.e. 0.71 = 71%)
G = initial gravity (in °P)
I = Inoculation rate (in million per ml)

$$A = 9.54 \times 10^{-4}(G) - 2.44 \times 10^{-4}(G)(I) + 4.23 \times 10^{-3}(I) + 5.19 * 10^{-1}$$

Cell Growth as a Function of Sugar

In the early 1800s, Karl Balling made an observation about fermentation. What he saw then was just as true then as it is now. It's an equation that is often quoted by brewing chemists, and can be used to give a better cell count estimate than calculators based solely on initial conditions.

Hearing that there is a better way to estimate cells than using an online calculator is likely not a surprise for anyone who has counted cells produced by a starter. After performing a number of tests on starters I have concluded that the cell growth starters don't match the results of the calculators. Other people that I have talked to that count the cells produced by their starters see this as well.

Most online calculators use the initial cell count and the volume of wort to estimate the final cell count. What I have found is that cell growth is rarely a function of volume or initial cell population, but a function of consumed sugar. My observations roughly follow the old brewer's rule of 10 billion cells produced per liter per °P. I've struggled to find where this rule comes from. A well-respected brewing chemist quoted Balling's observation of normal fermentation:

2.0665g sucrose -> 1g Ethanol + 0.9565g CO_2 + 0.11g of other constitutes.

But instead of "other constitutes" he said "yeast." Of course this makes sense! The sucrose is used to store energy as ATP and NAHD in the yeast, create sterols in the yeast, and build cell walls... all in the yeast!

Converting the equation for 1 liter of 1°P wort we have:

10g sucrose -> 4.84g Ethanol + 4.63g CO_2 + 0.532g yeast

The dry weight of yeast is 20 billion cells per gram. 0.532 * 20 = 10.6 billion cells!

Or put another way, for every gram of sugar consumed there are 1 billion cells produced.

It really is this easy:

1g fermentable extract = 1billion cells.

For a starter add about 20-30% extra malt to compensate for the unfermentable sugars. A refractometer is a wonderful tool for measuring the composition of small worts such as starters. Using an initial gravity and final gravity reading you will be able to estimate the number of cells grown in the wort at any given time. When measuring with a refractometer the sugar content needs to be compensated for alcohol.[20] This can be worked into the equation:

OGR = Original Gravity measured with a Refractometer.
FGR = Final Gravity measured with a Refractometer.
Number of cells per liter (in billions)=(OGR-(1.128*FGR-0.128OGR))*10

The popular online calculators are useful to get a starter set up. Initial and final gravity readings can be taken to get a much better idea of how many cells are produced. When the desired cells have been grown the starter can be refrigerated to stop cell production and separate the yeast from the spent wort. As an alternative, the fermentation can be allowed to run to completion. Then the cell count can be estimated and the correct portion pitched for the beer.

This accounts for what Balling describes as normal fermentation. There are other factors than the amount of sugar in the initial wort. Unmetabolized sugar will not be converted to yeast. Oxygen can be a limiting factor, but if your pitch rate is between 10 and 100 million cells per ml and you give the container a good shake when pitching things should work. Alcohol generated can kill yeast so the gravity should be

---

[20] See Chapter on Refractometer corrections for alcohol.

kept below 10°P (1.040) or about 100g per liter of water. (1/4c DME in 1 quart)

°B	OG SG	Final Gravity (°B)										
		1	2	3	4	5	6	7	8	9	10	11
1	1.004	0										
2	1.008	11	0									
3	1.012	23	11	0								
4	1.016	34	23	11	0							
5	1.020	45	34	23	11	0						
6	1.024	56	45	34	23	11	0					
7	1.028	68	56	45	34	23	11	0				
8	1.032	79	68	56	45	34	23	11	0			
9	1.036	90	79	68	56	45	34	23	11	0		
10	1.040	102	90	79	68	56	45	34	23	11	0	
11	1.044	113	102	90	79	68	56	45	34	23	11	0
12	1.049		113	102	90	79	68	56	45	34	23	11
13	1.053		124	113	102	90	79	68	56	45	34	23
14	1.057		135	124	113	102	90	79	68	56	45	34
15	1.061		147	135	124	113	102	90	79	68	56	45
16	1.066		158	147	135	124	113	102	90	79	68	56
17	1.070		169	158	147	135	124	113	102	90	79	68
18	1.075			169	158	147	135	124	113	102	90	79
19	1.079			180	169	158	147	135	124	113	102	90
20	1.084			192	180	169	158	147	135	124	113	102
21	1.088			203	192	180	169	158	147	135	124	113

Table 18 - Cell Growth from Refractometer Measurements

Billion cells per liter.

The columns listed across the top are the final gravity as read by a refractometer. These numbers are compensated for alcohol.

The rows listed down the left side are the original gravity as read by a refractometer.

Yeast Pitching

6

Starter Cell Count Estimators

Yeast pitch rates can greatly affect the beer produced. Even a factor of two can be significant.[21] There are several calculators available that can be used to determine the amount of slurry to pitch, and the volume of a starter.[22] [23] At first sight these calculators appear to be the perfect answer for pitching cells from a starter. Numerous factors are entered into the equation. The harvest date of the yeast, the thickness of slurry, and several of other factors are applied. However the results may not be nearly as accurate as you may hope. The first (and last) time I used one of these calculators for slurry viability the calculator indicated that in order to pitch the 125 billion cells for the beer I was brewing I would need 7.5 ounces of my slurry.

Later I would find that a cell count showed that the calculator I used was off by more than a factor of 10!

The beer that yeast went into was a Saison and it finished disappointingly with none of the recognizable yeast character that is typical for the style. Most of the error can be attributed to the viability by date portion of the calculator. Time in the refrigerator should not be counted when assessing viability. (More on this later.)

About a week after this mistake I purchased a microscope and performed several cell counts on some of the remaining slurry. The results were vastly different from the estimate from the calculator. The slurry that was calculated as 16% viable based on the collection date was actually 60% viable, and the cell density was not 400 billion cells per liter, but rather 2000 billion cells per liter. The result was that I had pitched almost twenty times more cells than were required. As you

---

[21] http://www.wyeastlab.com/hb_pitchrates.cfm
[22] http://www.mrmalty.com/calc/calc.html
[23] http://yeastcalc.com/

might expect the fermentation process was ridiculously fast and it reached the final gravity in less than 24 hours.

With a microscope and a hemocytometer, it is a pretty simple task to count viability. My first time counting cells went pretty quickly. We measured viability and counted cell density of four slurries in about an hour. Since then I have done hundreds of cell counts and reduced the time to about five minutes. The best way to know how much yeast is being pitched is to count the cells. (See the Brew Lab Technique section for details on how to count cells.)

Rehydrating Yeast

Dry yeast is a very viable alternative to liquid yeast... no pun intended.

Yeast do not hydrate effectively in wort. When the yeast cell is properly hydrated the cell membrane can regulate the amount of sugar it absorbs. However during hydration the membrane is ineffective. The yeast cell is filled with more sugar than it can process, systems inside the cell fail to function and the cell dies.

Fermentis recommends hydrating yeast at 80°F (27°C).[24] One study shows that viability of rehydrated yeast roughly follows a linear relationship to temperature.[25] Viability was 2% at 50°F (9°C) and 65% at 125°F (50°C). This indicates that viability will be only 20% at typical ale temperatures and 5% at lager temperatures! Experiments by a fellow home brewer seem to confirm this.[26] Another set of experiments seems to show that temperature has little effect of viability.[27] I could conduct yet another experiment but given the lack of correlation from the other experiments, even I would doubt the outcome would apply well in all circumstances.

However, all of the experiments and dry yeast manufacturers do agree on one thing: Viability will be high when yeast is rehydrated using 80°F (27°C) plain water and the yeast is allowed to float on the surface for 15 to 30 minutes.

There is an easy way to accomplish this. In a sanitized container microwave two cups of room temperature distilled water for 1 minute on high. This should bring the water to just over 100°F. Pour this into an ale-pail and the temperature will drop to about 85°F. You may want to measure the temperature and adjust the time if needed. Sprinkle the

---

[24] http://www.fermentis.com/wp-content/uploads/2012/02/SFA_S04.pdf
[25] http://onlinelibrary.wiley.com/doi/10.1046/j.1365-2672.1999.00638.x/pdf
[26] http://bkyeast.files.wordpress.com/2013/03/meschart.jpg
[27] http://seanterrill.com/2011/04/01/dry-yeast-viability/

yeast onto the surface of the water from just above the surface. The less distance the yeast fall the better their chance of floating. Even if the water doesn't entirely cover the surface of the pail don't worry: studies show that slowly rehydrating yeast is more effective than rapid rehydration.[28] [29] As long as the yeast is damp it will have a good chance of proper hydration. After 30 minutes the temperature will drop to about 80°F which allows plenty of margin to stay within the manufacturer's recommended temperature for the entire rehydration time.

As long as the temperature is close the yeast should rehydrate well. The temperature doesn't affect the viability as much as I thought it would. In a test at four different temperatures from about 70°F to 90°F the viability was about the same across the samples. Average temperature was calculated by integrating the temperature over time.

While yeast is on the surface of the liquid it is in contact with the oxygen in the air. This allows for aerobic respiration which is much more efficient than anerobic respiration. As yeast is hydrating this is critical. As the yeast absorbs the water it gets a little heavier. At some point it reaches a tipping point where it sinks beneath the surface of the water. Once it is below the surface the oxygen is, of course, extremely restricted.

In the experimental cases the lowest viability was seen in the yeast that sank first, and the highest viability was in the container in which the yeast sank last.

From these results it seems that using a container with a large surface area is important for hydrating yeast.

---

[28] Kosanke, J.W., Osburn, R.M., Schuppe, G.I. and Smith, R.S. (1991) Slow rehydration improves the recovery of dried bacterial populations. *Canadian Journal of Microbiology* **38**, 520–525.
[29] Leach, R.H. and Scott, W.J. (1959) The influence of rehydration on the viability of dried microorganisms. *Journal of General Microbiology* **21**, 295–307

How Many Cells in a Package

White Labs and Wyeast advertise 100 billion cells per container. Fermentis advertises a minimum of 6 billion cells per gram (69 billion total cells) at time of packaging.

To better understand the yeast pitched, over the last several weeks I have been doing cell counts on yeast purchased from the local home brew store.

Number	Strain	Viable cell count	Viability
WLP650	Brettanomyces Bruxellensis	102 billion	64%
WLP862	Cry Havoc	93 billion	98%
US-05	American Ale	77 billion *	30%
T-58	Safbrew Belgium	153 billion **	74%
WB-06	Safbrew Wheat	260 billion **	74%

Table 19 - Cell Counts of Various Packages

* rehydrated in cold water for about 5 minutes, then stirred in.

** rehydrated in 77°F water for 30 minutes, then stirred in.

These three yeasts cover a very wide range: both liquid and dry, brewers and wild strains. There was very little actual content in the WLP650 vial compared to what you might expect with a typical brewers strain. Microscope examination showed that the cellar morphology is very different. The WLP862 looks like a typical brewer's yeast: spherical and 15-20 micrometers in diameter. The WLP650 was football shaped, although there was some variation in cell shape. They also appear to be much smaller at about 7 microns in diameter and about 10 microns long.

The US-05 looks like typical brewer's yeast. I was surprised that the viability was so low. This might be due to the rehydration method deviating from the manufacturer's recommendations. This is not a problem, though, as the viable cell count is high enough.

Yeast Pitching Rates

Calculating the number of cells to pitch doesn't need to be complicated. From measurements done on yeast from a variety of manufacturers the cell count can deviate by 20%.[30] Propagation of yeast is also hard to predict. With a cytometer and a very careful count higher accuracy can be achieved, but for beer a 50% error pitch rate will likely go unnoticed. The pitch rate recommended by Wyeast jumps by a factor of two at 15 degrees Plato.[31] Pitch rates are important, but there is considerable leeway. However, it doesn't hurt to be as accurate as possible because compounding errors can quickly get the cell count outside of reasonable limits.

For Ales 750 million cells per liter per degree Plato, and for lagers 1500 million cells per liter is widely accepted.[32] The most commonly available yeasts come in packages containing either 100 billion or 150 billion cells.

Volume		\multicolumn{10}{c}{Original Gravity (Specific Gravity)}
gal	L	1.020	1.030	1.040	1.050	1.060	1.070	1.080	1.090	1.100	1.110	1.120
2	8	50	50	50	50	100	100	100	150	150	150	150
3	11	50	50	100	100	150	150	150	200	200	250	250
4	15	50	100	100	150	150	200	250	250	300	300	350
5	19	50	100	150	200	200	250	300	300	350	400	450
6	23	100	150	150	200	250	300	350	400	450	450	500
7	26	100	150	200	250	300	350	400	450	500	550	600
8	30	100	150	250	300	350	400	450	500	550	600	700
9	34	150	200	250	300	400	450	500	550	650	700	750
10	38	150	200	300	350	450	500	550	650	700	800	850
11	42	150	250	300	400	450	550	600	700	800	850	950
12	45	150	250	350	450	500	600	700	750	850	950	1000
		5	8	10	13	15	18	20	23	25	28	30
		\multicolumn{11}{c}{Original Gravity (°P)}

Table 20 - Ale Pitch Rate

---

[30] See Previous Section "How Many Cells in a Package"
[31] http://www.wyeastlab.com/com-pitch-rates.cfm
[32] http://www.whitelabs.com/beer/homebrew_information.html

Reduce the Wort Instead of Increasing the Yeast.

Sure, making a starter is cheaper than buying multiple vials of yeast, but the accuracy, scheduling, and waste of fermentables bothers me. However, it may seem like a necessity to better approximate the pitch rate required if you don't have a microscope. I don't think anyone will disagree when I say that yeast is a major cost contributor to the beer.

So what's a better way to ensure the correct cell count for your beer?

For a 5-gallon batch of standard ale with an initial gravity of 1.060, 200 billion cells are required. There are several ways this can be achieved.

1) Buy two yeast vials (2 x $8 = $16).
2) Buy one yeast vial and use a 2 liter starter ($8 + ~$2.00 = $10).
3) Buy one yeast vial and pitch into less wort ($8).

With the third method listed the cost of yeast is half of what it could have been! Starters themselves are not as accurate as they may seem. Sure Yeast Calc and Mr. Malty give you three digits of precision, but I would be surprised if they are within 25% of the cell count predicted with a wide variety of yeast strains.

When pitching yeast, the important thing is to have the correct number of cells per volume of wort. Instead of increasing the number of cells needed for the wort, the amount of wort can be reduced to the appropriate volume for the number of cells you have. For the 5 gallon 1.060 case here, one vial of yeast has half the cells we need, so these should be pitched into 2.5 gallons of the wort. The following day, once fermentation has picked up, the remainder of the wort can be added. Viable cell count can double within a few hours of the start of cell division.

Viability of the yeast you pitch may be a concern. Sure, when the yeast left White Labs for your local home brew store it had very close to 100 billion cells, but what is the condition when you picked it up for your brew? Tests I have conducted indicate that loss of viability is very slow with refrigerated yeast. In fact, I have seen no noticeable drop in viability over the course of a month of the slurries in my refrigerator. I would imagine that professionally packaged yeast would only perform better.

One concern you may have with this method is infection. Because a large percentage of the wort remains for 24 hours without being inoculated the chance that something else may get a head start is a possibility. However, if you make sure to sanitize everything, there should be no issue with holding the wort for a day before pitching.

On brew day sanitize two fermenters. Chill the wort and pour half into each one. Pitch the yeast into your primary fermentation vessel, and seal both containers. The following day, just like a starter, your yeast will have grown significantly. Pour the wort from the second container into the primary fermentation vessel. For a bonus, pour as vigorously as you like. Aeration is beneficial at this stage of fermentation.

An alternative to a second pail would be to keep the wort in your boil kettle and seal it up. My boil kettle is a pressure caner, so this is an easy task. If you are using a traditional kettle you may need to get creative. An alternative would be to pour the hot wort into a large expandable water container as done in the no-chill method.

It's that easy. No starter to make or clean. No delicate scheduling to ensure you have enough cells on brew day. There is only one more thing to sanitize which should take an insignificant amount of time when done in parallel to sanitizing you primary fermentation vessel.

Do the Two Step

This is the most foolproof way to pitch yeast that you will likely ever hear.

Without a microscope you are guessing at how many cells you are pitching. Over the last few months you've seen experimental results that show that slurry viability varies greatly and the density is far from predictable.

There are brewer's rules of thumb that have stood the test of time, and the combination of these two produce an eloquent way to pitch the perfect amount of yeast into your beer.

A) Standard Pitch Rate is 1 billion cells per Liter per degree Plato.
B) Cell growth is 10 billion cells per Liter per degree Plato.

When these two rules are next to each other what they have in common becomes vibrantly clear. They are both a function of liters per degree Plato. The cells grown by the starter need to equal the cells pitched for the beer. Because both worts have the same gravity, the gravity drops out of the equation, leaving a very simple relationship:

To grow the correct number of cells the starter should be 1/10 of the wort. That's it.

So for a five-gallon batch you would use a ½ gallon starter of the same wort you would pitch into. One half gallon of wort is much easier to cool than the entire five gallons making it possible to pitch the yeast on brew day without an immersion chiller or several sinks full of ice.

This can also be applied to pitching on top of a cake. Because the cell growth is a factor of ten during a regular fermentation only 1/10th of the yeast cake is needed for the next beer if it is the same gravity. The viability of the yeast also needs to be considered. From earlier experiments it has been demonstrated that alcohol in the beer will affect yeast viability.[33] To prevent under pitching more than 1/10th of the cake should be used.

1/5th of the cake should be used for a new beer of equal gravity.

I'll hash out the math:

PCC = Pitching Cell Count (in billions)
OG = Original Gravity (in degrees Plato)
V = Volume of beer wort (in Liters)
SV = Starter Volume (in Liters)
SC = Starter Cells generated (in billions)

Equation A: PCC = OG * V
Equation B: SC = 10 * SV * OG
OG * V = 10 * SV * OG
1/10 V = SV

This new method has several distinct advantages to other pitching methods I have tried.

1) The cells pitched into the beer are recently propagated making them healthy and highly viable.

2) The number of cells is not dependent on the viability, vitality, or number of yeast cells in the starter

3) There is zero math or calculators involved.

---

[33] See "ABV effects on Yeast" Section

Aeration

Without a DO meter we are all guessing as to how much oxygen has been added to our wort at the start of fermentation. Oxygen is important in the initial phase of fermentation. It is required in the production of sterols which allow for cell wall permeability. Without a sufficiently permeable cell wall sugars cannot flow freely into the yeast. This makes it difficult for the yeast to obtain the nutrients they need and will leave the yeast in poor health. The yeast will not be able to divide to the population creating the same effects as under pitched yeast. In addition the fermentation will be slow, and the yeast will be stressed which can result in stress excretions by the yeast. [34] [35] All around under oxygenating is bad news.

This leads to the question of "can you have too much oxygen?" The simple answer is yes. Fix goes into detail about the repercussions in "Principles of Brew Science."[36] If you haven't read this book; I highly recommend that you pick up a copy. For starters, excessive oxygenation leads to excessive cell division which leads to similar problems with over pitching.[37] Over oxygenating is a good way to get low phenol and low ester aromas and flavor out of your yeast. For a lager this might be a good thing, but for most Belgium, German, and English styles you'll want that contribution.

True saturation is never really possible, but 95%-99% should be achievable and repeatable. Some tests show that rocking and shaking can achieve this level in minutes.[38] Additionally, the saturation point could change about 5% based on other factors such as temperature, pressure, and to a small degree, wort content. The saturation point of

---

[34] http://www.wyeastlab.com/hb_oxygenation.cfm
[35] http://www.howtobrew.com/section1/chapter6-9-2.html
[36] "Principles of Brewing Science" Fix
[37] http://www.wyeastlab.com/hb_pitchrates.cfm
[38] http://www.brewangels.com/Beerformation/AerationMethods.pdf

sucrose is 2000 grams per liter.[39] Compare this to a 1.100 wort (for a very big beer) that would only have about 280 grams of sugar per liter.

With air, the saturation point of oxygen is 8ppm. No matter how much you mix, pour, shake, and pump air into the wort, it's not getting any higher than 8ppm. This can be used to your advantage. If you want 4ppm, or half of the saturated content, then aerate half of your wort, and then mix that with the other half that has not been aerated.

With pure oxygen the saturation point is about 50ppm. The sample principles can be applied to using pure O2. If an ale-pail is purged with oxygen, and then agitated, it can be saturated to 50ppm of oxygen. This is much more oxygen than is needed for any reasonable beer.

The table below shows the percentage of the wort that will need to be saturated to achieve the desired oxygen content in the wort.

Gravity		Oxygen required	Saturate with	Percentage to saturate	Gallons to Saturate
SG	°P				
1.008	2	2	air	25%	1.25
1.016	4	4	air	50%	2.5
1.024	6	6	air	75%	3.75
1.032	8	8	air	100%	5
1.040	10	10	O_2	20%	1
1.049	12	12	O_2	24%	1.2
1.057	14	14	O_2	28%	1.4
1.066	16	16	O_2	32%	1.6
1.075	18	18	O_2	36%	1.8
1.084	20	20	O_2	40%	2
1.093	22	22	O_2	44%	2.2

Table 21 - Aeration

---

[39] http://en.wikipedia.org/wiki/Sucrose

Pitch Rates and Starters

Some brewers, including myself, enjoy making a science project out of yeast propagation, but it certainly doesn't have to be. In a busy life style even enjoyable tasks become chores when crushed in the vice of time. Under this pressure comes the need to save time without reducing quality. Simplifying the yeast propagation process becomes a necessity.

As we have seen, there are a number of ways to achieve the correct pitch rate without creating a starter. However, there may be an instance where growing yeast is preferred.

Yeast propagation can be an elaborate process using laboratory equipment such as a flask and a stir plate. It certainly looks sophisticated. It may produce more yeast faster, but it doesn't make better yeast.

First, to propagate yeast, a vessel to contain the fermentation is needed. Nearly any container that can be sanitized and covered will work fine. A quart size glass canning jar works well for propagating up to 100 billion cells. To propagate more cells a plastic food storage container, gallon glass wine bottle or buckets will suffice. If you chose plastic for propagating yeast it will need to be free from damage. If scraped, even a thoroughly washed and sanitized plastic container can harbor bacteria.

Next, yeast must have food. Dried malt extract has been proven to be a wonderful source of nutrients for yeast. It contains not only sugars, but also many of the minerals that the organisms require to thrive. The amount of extract used is directly proportional to the number of cells produced. Table sugar may seem like a more economical solution, but the results are not well tested. (See the Balling Observation.)

Finally, the yeast needs to be considered. Cell density does have some effect on propagation, but not as much as some of the online calculators may lead you to believe. Extensive testing done by other brewers as well as myself show that pitch rate has very little effect on the number of cells grown. (See Yeast Extrapolated Data.) A pitch rate between 1 billion and 100 billion cells per liter is preferred to produce predictable results. The highest density would be one vial, smack pack, or package of yeast in one liter of wort. The Lowest density is roughly one teaspoon of slurry in five liters of wort.

Once everything is combined the starter should be aerated and then covered. Aluminum foil will prevent dust from falling in but allow sufficient air flow for the carbon dioxide to escape and oxygen to enter the fermentation vessel. Yeast propagation can take as little as 24 hours for a starter with a high pitch rate to almost a week for starters with a low pitch rate.

BREWING ENGINEERING

gallons

°P	SG	2	3	4	5	6	7
7	1.028	-	-	-	-	-	-
8	1.032	-	-	-	-	-	50g, 1L
9	1.036	-	-	-	-	50g, 1L	100g, 1L
10	1.040	-	-	-	50g, 1L	100g, 1L	100g, 1L
11	1.044	-	-	-	50g, 1L	100g, 1L	150g, 1L
12	1.048	-	-	-	100g, 1L	100g, 1L	150g, 2L
13	1.053	-	-	50g, 1L	100g, 1L	150g, 1L	150g, 2L
14	1.057	-	-	50g, 1L	100g, 1L	150g, 2L	200g, 2L
15	1.061	-	-	100g, 1L	150g, 1L	150g, 2L	200g, 2L
16	1.065	-	-	100g, 1L	150g, 1L	200g, 2L	250g, 2L
17	1.070	-	50g, 1L	100g, 1L	150g, 2L	200g, 2L	250g, 3L
18	1.074	-	50g, 1L	100g, 1L	150g, 2L	200g, 2L	300g, 3L
19	1.079	-	50g, 1L	150g, 1L	200g, 2L	250g, 2L	300g, 3L
20	1.083	-	100g, 1L	150g, 1L	200g, 2L	250g, 3L	300g, 3L
21	1.087	-	100g, 1L	150g, 2L	200g, 2L	300g, 3L	350g, 3L
22	1.092	-	100g, 1L	150g, 2L	250g, 2L	300g, 3L	350g, 4L
23	1.096	-	100g, 1L	200g, 2L	250g, 2L	300g, 3L	400g, 4L
24	1.101	-	100g, 1L	200g, 2L	250g, 3L	350g, 3L	400g, 4L
25	1.106	50g, 1L	150g, 1L	200g, 2L	300g, 3L	350g, 4L	450g, 4L
26	1.110	50g, 1L	150g, 1L	200g, 2L	300g, 3L	350g, 4L	450g, 4L
27	1.115	50g, 1L	150g, 1L	200g, 2L	300g, 3L	400g, 4L	450g, 5L
28	1.120	50g, 1L	150g, 2L	250g, 2L	300g, 3L	400g, 4L	500g, 5L
29	1.124	50g, 1L	150g, 2L	250g, 2L	350g, 3L	400g, 4L	500g, 5L
		8	12	16	20	24	28

liters

Table 22 - Starter Size from Packaged Yeast

When propagating yeast from a package the initial cell population will be about 100 billion cells. Depending on the original gravity and volume of the beer propagation may not be necessary. These conditions are shown as "-" in the table. For beer where a starter is recommended the weight of sugar is shown in grams and the volume of water is shown in liters. The volume of water is much less critical than the amount of sugar. If the vessel is approximately the correct size no measuring is needed.

One-hundred grams of dried malt extract is roughly one quarter of a cup. One liter is roughly one quart.

Making Great Beer Through Applied Science

		gallons						
°P	SG	1	2	3	4	5	6	7
1	1.004	50g, 1L	50g, 1L	50g, 1L	50g, 1L	50g, 1L	50g, 1L	50g, 1L
2	1.008	50g, 1L	50g, 1L	50g, 1L	50g, 1L	100g, 1L	100g, 1L	100g, 1L
3	1.012	50g, 1L	50g, 1L	100g, 1L	100g, 1L	100g, 1L	100g, 1L	100g, 1L
4	1.016	50g, 1L	50g, 1L	100g, 1L	100g, 1L	100g, 1L	100g, 1L	150g, 1L
5	1.020	50g, 1L	100g, 1L	100g, 1L	100g, 1L	150g, 1L	150g, 1L	150g, 2L
6	1.024	50g, 1L	100g, 1L	100g, 1L	100g, 1L	150g, 1L	150g, 2L	200g, 2L
7	1.028	50g, 1L	100g, 1L	100g, 1L	150g, 1L	150g, 2L	200g, 2L	200g, 2L
8	1.032	50g, 1L	100g, 1L	100g, 1L	150g, 1L	150g, 2L	200g, 2L	200g, 2L
9	1.036	100g, 1L	100g, 1L	150g, 1L	150g, 2L	200g, 2L	200g, 2L	250g, 2L
10	1.040	100g, 1L	100g, 1L	150g, 1L	150g, 2L	200g, 2L	250g, 2L	250g, 3L
11	1.044	100g, 1L	100g, 1L	150g, 1L	200g, 2L	200g, 2L	250g, 2L	300g, 3L
12	1.048	100g, 1L	100g, 1L	150g, 2L	200g, 2L	250g, 2L	250g, 3L	300g, 3L
13	1.053	100g, 1L	150g, 1L	150g, 2L	200g, 2L	250g, 2L	300g, 3L	300g, 3L
14	1.057	100g, 1L	150g, 1L	200g, 2L	200g, 2L	250g, 3L	300g, 3L	350g, 3L
15	1.061	100g, 1L	150g, 1L	200g, 2L	250g, 2L	300g, 3L	300g, 3L	350g, 4L
16	1.065	100g, 1L	150g, 1L	200g, 2L	250g, 2L	300g, 3L	350g, 3L	400g, 4L
17	1.070	100g, 1L	150g, 2L	200g, 2L	250g, 3L	300g, 3L	350g, 4L	400g, 4L
18	1.074	100g, 1L	150g, 2L	200g, 2L	250g, 3L	300g, 3L	350g, 4L	450g, 4L
19	1.079	100g, 1L	150g, 2L	200g, 2L	300g, 3L	350g, 3L	400g, 4L	450g, 4L
20	1.083	100g, 1L	150g, 2L	250g, 2L	300g, 3L	350g, 4L	400g, 4L	450g, 5L
21	1.087	100g, 1L	200g, 2L	250g, 2L	300g, 3L	350g, 4L	450g, 4L	500g, 5L
22	1.092	100g, 1L	200g, 2L	250g, 2L	300g, 3L	400g, 4L	450g, 4L	500g, 5L
23	1.096	100g, 1L	200g, 2L	250g, 3L	350g, 3L	400g, 4L	450g, 5L	550g, 5L
24	1.101	100g, 1L	200g, 2L	250g, 3L	350g, 3L	400g, 4L	500g, 5L	550g, 6L
25	1.106	150g, 1L	200g, 2L	300g, 3L	350g, 4L	450g, 4L	500g, 5L	600g, 6L
26	1.110	150g, 1L	200g, 2L	300g, 3L	350g, 4L	450g, 4L	500g, 5L	600g, 6L
27	1.115	150g, 1L	200g, 2L	300g, 3L	350g, 4L	450g, 5L	550g, 5L	600g, 6L
		4	8	12	16	20	24	28
		liters						

Table 23 - Starter size for Slurry Yeast

Yeast, when propagated from a small amount of slurry, has a very small initial population. This table shows the size of a starter required to generate all of the cells recommended for beers of varying sizes.

Storing Yeast

7

Yeast "Washing"

The definitions of yeast washing differ between that used in the professional brewing industry and by the home brewer. In the brewing industry yeast washing is a method to kill unwanted bacteria that may be mixed in with the yeast just prior to pitching. Most home brewers use the term to describe a process used to rinse unwanted particulate from a yeast slurry before storage.

In home brew yeast washing, or more accurately, yeast rinsing, the yeast is harvested from a fermentation vessel and combined with water. It is allowed to settle until the water has separated to the top and the yeast to the bottom. The water and the top of the slurry is then poured into several jars. The yeast and debris that has settled to the bottom of the first container is then discarded.

Normally I don't wash my yeast, I just pour the cake into three or four quart size mason jars, but I have been curious as to what benefit may be achieved from yeast washing. You have probably heard that the intent is to separate the hops and dead yeast cells from the live yeast cells. Over the course of several months I have run experiments on this technique to find out what it does, and if it is worth the extra time.

Yeast washing can be beneficial, but not for the reasons that I had anticipated.

After the yeast has settled into the container it divides into roughly three sections. Common brewing wisdom indicates that the top portion is mostly water, the light colored middle section contains viable yeast, and the darker bottom contains dead yeast, hops, and other debris. However, it seems that this is not the case.

The viability throughout the container is roughly the same.

In four test cases the viability was not statistically different in these layers. Tests were run with three slurries with 10%, 50% and 90%

viability. In all tests the viability of the yeast in each section did not vary more than one standard deviation.

What was interesting was that the bacteriological content was much higher in the top portion of the yeast containers than in the lower parts. There was about 100 times more bacteria per live yeast cell in the top "liquid" section.

Another strange finding was that the concentration of non-yeast debris followed the cell density. While the cell concentration at the bottom of the container was twice what it was in the middle, the viability was the same, and the concentration of non-yeast material per yeast cell was virtual identical. The hops and other partials did not separate from the yeast. So if this "junk" is discarded, it takes just about as much viable yeast with it as it takes debris.

The first set of experiments was done in test tubes. To make sure the results were repeatable; yeast washing was performed with a full sized batch. The results were the same. The viability of the part that would have been thrown away was the same as that of the part that would have been kept. The viable cell density was also very similar.

The tests were conducted with WLP566, WLP004, EC-1118 and S-04.

Conclusion

When washing yeast, discard the liquid to remove bacteria. Keep the thick slurry and add clean water on top of it for storage and to wash out additional bacteria.

Rinsing Yeast From a Fruit Beer

Yeast rinsing did not successfully isolate the viable cells, but for completeness I decided to wash the entire slurry from a beer made with fruit to really allow the yeast washing to work. Although the washing didn't effectively separate live cells from dead cells perhaps it would be suitable for separating fruit partials from yeast cells.

Yeast washing, or more accurately yeast rinsing, procedure:

1) Add water to the slurry and mix. Wait 20 minutes.
2) Pour off the top part of the slurry that hasn't settled. Allow this to settle for 20 minutes.
3) Pour off the top of the slurry into jars to save the "washed" yeast.

The result is much less yeast that you started with.

Pour	Viability	Cell volume	Cell count
First discarded yeast (pour 1-2)	20%	250ml	11 billion
Second discarded yeast (pour 2-2)	26%	30ml	470 million
Retained yeast (pour 2-1)	31%	15ml	470 million

Table 24 - Yeast Washing Cell Counts

Note that the total cell count was about 12 billion cells, but following the yeast rinsing procedure less than 500 million cells were saved. The viability of all of the layers was approximately the same.

In summary, yeast washing was successful at separating the cellulose from the yeast, but at the cost of losing an enormous amount of good yeast (95% of the total cells produced). If you just put the yeast into jars from the fermenter you will have 25 times more yeast that is about as viable as the water washed yeast.

Yeast Storage

Maintaining yeast over long periods of time is often over-thought resulting in more chances of contamination and loss of 95% of the viable yeast. Water washing of yeast, often referred to as yeast washing or yeast rinsing, results in discarding all but a one twentieth of the viable yeast and, in my opinion, should not be done to prepare the yeast for storage.

The addition of fruit to the beer will affect viability.[40] Because the addition of fruit to the beer reduces viability the yeast should be harvested before adding the fruit. To maintain the health of the yeast, add the fruit to a secondary fermentation vessel and rack the beer on top of it. Then harvest the yeast from the primary vessel.

Storing yeast is simple.

1) After racking the beer off of the cake, slosh the vessels to homogenize the slurry.
2) Pour the slurry into three or four quart size mason jars.
3) Store the Jars in the refrigerator.

During the first few days of storage CO_2 will be released from solution. It is important that this gas can escape. The easiest way to accomplish this is to not tighten the lid all the way. Leaving it just a little tight for the first few days will allow the CO_2 to escape slowly and the positive pressure in the container will keep most contaminates from entering. If you are concerned then another option is to vent the containers a couple of times a day until all of the pressure has been released.

Yeast, when properly stored, will last for six months according to "The Practical Brewer"[41]

---

[40] See "How Fruit Effects Viability" Section
[41] The Practical Brewer - Yeast - Strains and Handling Techniques. p276 of the third addition.

Each jar will typically contain 300 billion to 2 trillion cells. If you don't have a microscope 1 trillion cells per quart is a good estimate. If the ABV of the beer that it was taken from was less than 6%, and there was no fruit added before the slurry was collected then it is fairly safe to assume the viability is 90%.[42] Viability does not drop much over time, so don't use any date-based calculations for viability. The type of beer that the slurry was taken from will have a much larger effect on viability.

---

[42] See "Refrigeration Effects on Yeast Viability" Section

Refrigeration and Viability

Some brewing-related subjects have a plethora of information available allowing the home brewer to evaluate different pieces of data and come to their own conclusion. However, other subjects may have one scrap of fact that is misused creating misinformation. The effects of refrigeration on viability are an example of the latter. Mr. Malty's slurry viability slider is often misrepresented in this manner. For new slurries this slider shows the viability as 94% dropping 1.6% every day, and bottoming out at 10% after 53 days. While Jamil likely had good intentions when designing the calculator, it is commonly misused. Perhaps it was an attempt to show that slurry cannot be kept in the refrigerator indefinitely. There are a number of other considerations when deciding to re-pitch slurry such as contamination and overall health of the yeast.

The fact is that it is documented in very reliable brewing literature that *yeast stored in a broth can be kept for six months, and yeast stored on an agar slant can be kept for over a year.*[43] [44]

There are much more important considerations when it comes to the viability of a starter than the amount of time it has been refrigerated. Fruit[45] has a drastic impact on viability as does alcohol.[46]

The linear decay has actually been propagated over at Yeast Calc, another trusted source. Rather than simply propagate information, I did the tests for myself. Over the course of a month data was collected on seven different slurries from two different strains used to make a variety of beers.

---

[43] The Practical Brewer, Yeast Strains and handling techniques, Sources and Maintenance of Pure Yeast Cultures. p276
[44] Kirsop (1991)
[45] See "How Fruit Effects Yeast Viability" Section
[46] See "ABV Effects on Yeast" Section

Yeast Strain	Yeast Layer	Viability loss per day
WLP566	Top	-0.07%
WLP004	Top	+0.006%
WLP004	Middle	-0.18%
WLP004	Bottom	-0.77%
WLP566	Top	-0.46%
WLP566	Top	-0.006%
WLP566	Bottom	+0.09%
Average		**-0.20%**

Table 25 - Viability Over Time

Conclusion:

While the initial viability can vary greatly, the viability over time does not change a measurable amount over the course of one month.

Yeast at Ambient

Without food yeast will begin to starve at ambient temperature conditions. This can lead to mutations in the yeast; however the number of living cells will stay the same for several days.

Even after three days outside of the refrigerator viability will not change more than 3%. The results from a viability test of yeast even shipped across the country should be fairly accurate.

When considering yeast health the first thing that comes to mind is viability: the percentage of live cells of the total cell population. However, there are other factors when considering yeast stored in the food danger zone. In this range of temperatures -- from 40 degrees F to 140 degrees F -- bacteria thrive.

Yeast that has been packaged for storage in a refrigerator is typically not stored with any food for the yeast because it is intended to stay dormant. When the yeast warms up to room temperature it will become active, and during this time the cell population will increase slightly, however it quickly runs out of energy reserves. The bacteria, on the other hand, will feed on the dead yeast cells, and continue to propagate even after the yeast stop growing.

During a test of two strains of yeast, WLP004 and WLP566, it was noted that within the first 12 hours both strains grew in population. Initial inspection of the yeast when it was removed for the refrigerator showed no signs of bacteriological activity, but after only 12 hours pairs of bacteria were observed indicating cell growth.

When the yeast is pitched into an environment that it thrives in, such as typical wort, it can out perform the bacteria and the defects produced by the bacteria may go unnoticed. If the bacteria are given a head start, such as a day or two at ambient conditions, they have a good chance of dominating the flavor profile.

In conclusion, I wouldn't pitch wort with yeast that has been left out for an extended period of time, but the viability count and cell count should be very accurate even if the sample has been warm for several days.

ABV Effects on Yeast

When yeast sits in the bottom of a fermentor for a period of time, the alcohol will kill some of the yeast, but how much yeast is killed?

Yeast has a somewhat varying tolerance to alcohol, but it is not as black and white as I thought. For example, WLP862, Cry Havoc, is tolerant to 5-10% ABV, but in an experiment it took about two weeks to reduce the yeast to 50% viability.

The alcohol tolerance listed by yeast manufacturers is slightly different than what is presented here. Yeast manufacturers want to ensure that the yeast will complete fermentation of wort that reaches a specified alcohol level. This graph shows the percentage reduction in viable cells over time when exposed to alcohol. This is analogous to how you might expect yeast to survive in a fermentation vessel after primary fermentation has completed.

Being able to estimate the viability of the slurry based on time and alcohol content should give you a reasonable estimate of the viability of harvested slurry. In addition the amount of cell death might give you an idea of the amount of autolysis related flavors that may be added to your beer.

Five different mixtures containing the same amount of yeast, and varying levels of alcohol were allowed to sit at an ambient air temperature of approximately 60°F. Cells were stained with methylene blue and the number of live cells were counted. The graph shows the number of live cells compared to the initial number of live cells counted. Because there was no growth, the numbers above 100% will give you an idea of the error associated with the cell counts. No method of viability testing is perfect, but these results should get you in the ballpark.

Making Great Beer Through Applied Science

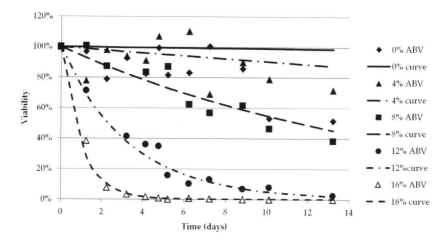

The top two sets of data, for 0% ABV and 4% ABV show very little drop in viability over time, but notice at about day ten there seems to be a significant drop. This corresponds to the time I also started noticing mold cells growing with the yeast. Unfortunately, daily measurements created dozens of opportunity for contamination.

Cell death was roughly exponential. At 16% ABV the death rate was approximately 70% death per day, whereas at 8% ABV the death rate was only 6% per day. This performance gives a window into what you might expect for viability cells removed from slurry. From the data collected here the death rate based on time and percentage of alcohol can be extrapolated.

Brewing Engineering

Alcohol by Volume

days	1%	2%	3%	4%	5%	6%	7%	9%	11%	13%	15%	17%
1	100%	100%	100%	99%	99%	98%	97%	94%	86%	70%	44%	15%
2	100%	99%	99%	98%	98%	96%	95%	88%	74%	49%	19%	2%
3	99%	99%	99%	98%	97%	95%	92%	82%	64%	35%	8%	0%
4	99%	99%	98%	97%	95%	93%	90%	77%	55%	24%	4%	0%
5	99%	98%	98%	96%	94%	91%	87%	73%	47%	17%	2%	0%
6	99%	98%	97%	96%	93%	90%	85%	68%	41%	12%	1%	0%
7	99%	98%	97%	95%	92%	88%	83%	64%	35%	9%	0%	0%
8	98%	97%	96%	94%	91%	87%	80%	60%	30%	6%	0%	0%
9	98%	97%	96%	93%	90%	85%	78%	56%	26%	4%	0%	0%
10	98%	97%	95%	93%	89%	84%	76%	53%	22%	3%	0%	0%
11	98%	96%	95%	92%	88%	82%	74%	49%	19%	2%	0%	0%
12	97%	96%	94%	91%	87%	81%	72%	46%	16%	1%	0%	0%
13	97%	96%	94%	91%	86%	79%	70%	43%	14%	1%	0%	0%
14	97%	96%	93%	90%	85%	78%	68%	41%	12%	1%	0%	0%
21	96%	93%	90%	85%	78%	69%	56%	26%	4%	0%	0%	0%
28	94%	91%	87%	81%	72%	61%	46%	17%	1%	0%	0%	0%
35	93%	89%	84%	76%	66%	53%	38%	11%	1%	0%	0%	0%
42	91%	87%	81%	72%	61%	47%	32%	7%	0%	0%	0%	0%
49	90%	85%	78%	69%	56%	42%	26%	4%	0%	0%	0%	0%
56	89%	83%	76%	65%	52%	37%	22%	3%	0%	0%	0%	0%
63	87%	81%	73%	62%	48%	32%	18%	2%	0%	0%	0%	0%
70	86%	80%	70%	58%	44%	28%	15%	1%	0%	0%	0%	0%

Table 26 - Viability of Yeast as a Function of ABV

Across the top, each column is for percent ABV of the finished beer.

Down the left, each row is the number of days at that ABV level stored at ambient.

$$p = 0.0014 * e^{0.4251*ABV}$$

$$viability = e^{-days*p}$$

Fruit and Viability

There are a number of factors you may have heard of that can reduce the viability of a yeast cake. These include, high gravity, high hops, and dark color, but one I have not heard of is the effect of fruit on viability. In the past few months I have made several fruit beers, and for three of them I collected viability data on the yeast slurries taken from these brews.

I have been surprised to find that yeast slurries from beer made with fruit have a drastically lower viability!

From one of these fruit beers I took a cell sample during fermentation just before adding the fruit and the viability was in the high nineties as you would probably expect during active fermentation. When the beer was bottled it was down to 1% viability.

The viability of the yeast that was taken from beers that did not have fruit has always been substantial higher. It looks like from this point forward, in order to save the yeast, I will be racking off of the yeast cake before adding fruit.

EC-1118 Champagne yeast harvested off of a Strawberry Champagne contained 19 billion suspended viable cells per liter during fermentation, but after adding the strawberries, the thick slurry only contained 6 billion viable cells per liter. There were three times as many viable cells suspend in the wine during fermentation as there were in the thick slurry after fermentation completed. This indicates that there was massive cell death after adding the strawberries.

Although this evidence seems fairly conclusive, I am hesitant to say this is always the case. But independent of the cause, there are significantly less viable yeast cells per liter of slurry. If a microscope was not used to asses viability before pitching, and a standard cell density was assumed, the wort would end up drastically under pitched, even if using a starter.

The fruit in these beers has pasteurized and has the cellulose removed with a strainer before being adding to the secondary fermentor. You can read exactly how it was done in the More Than Malt section. In the juice that is transferred to the fermentor there are likely still a plethora of fruit cells. It is possible that this large cellular mass is being counted as dead yeast cells and thus the viability shows as low.

Perhaps there is some merit to a secondary fermentation vessel.

Beer	Fruit	Yeast	Viability
Raspberry Cream	Raspberries	S-04	40%
Strawberry Champagne	Strawberries	EC-1118	1%
Saison Terri	none	WLP-566	87%
50c	none	WLP-566	85%
Raspberry Wheat	Raspberries	WLP-004	20%

Table 27 - Post Fermentation Viability

Real Yeast Washing

Yeast rinsing, as has been seen previously, does not adequately separate the dead cells, protein, or even trub from the yeast.[47] In fact the selection process of yeast rinsing encourages yeast that stay suspended. This means that the work you may be doing to select the best yeast is actually selecting yeast that makes a cloudy beer, and taking even more of the bacteria with it.[48]

Seven of the most common bacteria that we are fighting are:

Name	Defect	Morphology	Weakness
Acetic acid Bacteria	Produces vinegar	Alcohol and hop tolerant. Gram-negative, aerobic, rod-shaped bacteria (8)	
Acetomonas, Pseudomonadaceae	Apple cider smell	Alcohol and hop tolerant. Gram-negative, rod-shaped and polar-flagella (9)	
Coliforms	Celery and phenolic odors	rod-shaped Gram-negative(10)	
Lactobacillus	Produce Lactic acid	Gram positive 1µm by 5-120µm rod	Alcohol >5%
Obesumbacterium proteus	Parsnip flavors	Gram negative, 1.5-4µm rod	Hop, ethanol, pH
Pediococcus	Diacetyl, lactic acid	0.8-1.0µm balls in pairs or triads	
Zymomonas	Rotten apples	Rod	Hop, ethanol, pH

[49]

Table 28 - Summary of Bacterial Contaminates

---

[47] http://woodlandbrew.blogspot.com/2012/12/yeast-washing-exposed.html
[48] http://woodlandbrew.blogspot.com/2013/01/yeast-washing-revisited.html
[49] Principles of Brewing Science

Summary of Washing Techniques

Acid washing

This is the most common method used by the major breweries for removing bacteriological contamination from yeast.[50] It involves adding acid to lower the pH to 2 and allowing it to incubate for about an hour. The yeast is then pitched into the wort. Some major breweries use sulfuric acid, which can be extremely dangerous. At the home brew level Phosphoric acid may be suitable. Another option may be Acetic acid or vinegar although there would likely be detrimental impact to the flavor. The cost of washing a cup (250 ml) of yeast is $0.10.

Alcohol Washing

Most bacteria are sensitive to alcohol, including those that may have found their way into your slurry while it has been in the refrigerator. Yeast can tolerate up to 12% alcohol for 24 hours.[51] To wash a cup of yeast with alcohol will cost you about $2.00

Chlorine Dioxide

For home brewing this is one of the more common ways of washing yeast. It's the simplest of these methods as it's basically a one-step process. Major homebrew distributers even promote this method,[52] but the materials may be cheaper on Amazon. This will cost about $0.50 per wash.

---

[50] http://www.murphys-pro-system.com/datasheets/tech_wash.pdf
[51] http://woodlandbrew.blogspot.com/2013/01/abv-effects-on-yeast.html
[52] http://www.northernbrewer.com/shop/chlorine-dioxide-tablets-20-ct.html

Acid Washing

Using acid is no joke and requires proper handling. Many commercial breweries use sulfuric acid, although for home brewing this is likely not the best choice. Even Star San in its concentrated form, can be extremely caustic. Luckily for home brewers this procedure uses prepared Star San.

Star San, when properly prepared, is a 0.08% phosphoric acid solution. Think about the concentration: Even with just a trace of acid it is strong enough to kill nearly all bacteria. The pH of this solution is between 2 and 3 which is just about the recommended pH for yeast washing. If a small portion of yeast is combined with a much larger portion of Star San, the pH will stay in the yeast washing range. If this yeast slurry with acid was added to the beer it would likely be noticeably sour. To prevent this, the acid can be neutralized.

Phosphoric acid is classified as a strong acid, meaning that nearly all of the hydrogen ions disassociate. If this is combined with a small amount of a weak base the acid will pull all of the needed OH ions from the weak base till all of the hydrogen ions have been cancelled. Because the ions to not disassociate easily from the weak base creating a solution that is too alkaline is not a concern.

Another advantage of using a strong acid at a low concentration is that the concentration of salts produced by the reaction is very low.

One weak base that nearly everyone already has is baking soda, or sodium bicarbonate. Sodium bicarbonate by itself breaks down into carbon dioxide and sodium hydroxide. When combined with phosphoric acid three salts can be formed.[53] These are monosodium phosphate, disodium phosphate, and trisodium phosphate. These salts are fairly soluble in water, and will also be in low concentration.

---

[53] http://en.wikipedia.org/wiki/Sodium_phosphates

Crashing and decanting the yeast will remove a large percentage of these salts.

This theory was tested on slurry of yeast that showed low viability and contained easily identifiable bacteria. The initial viability was 42%, and the bacteria content was 1 bacterium for every hundred yeast cells. One culture was created with the washed yeast, and a second culture was made using the unwashed yeast. Other factors -- including aeration, original gravity, temperature and volume -- were identical.

Both cultures completed fermentation in less than three days. The attenuation was the same in both cultures. A carful microscope examination showed that there was no noticeable bacteriological growth in the washed yeast, but there was bacteriological growth in the yeast that was not washed.

Overall, using Star San to wash yeast is a viable option when working with yeast of questionable purity.

Home Brew Acid Washing

1) Combine 1 tablespoon (15 ml) of yeast with 1 cup (250 ml) of prepared Star Sans and allow to rest for one hour.
2) Add a pinch of baking soda to the solution and place in the refrigerator for at least 12 hours.
3) Decant the liquid, and use the remaining yeast for a starter.

Fermentation

8

How Often to Check Gravity

The common wisdom is that your beer has finished fermenting when two gravity readings are the same two or three days apart. While this is true for most beers it is not quite accurate for high gravity lagers which ferment over a long time. On the other hand, you might want to get a light bodied, highly hopped ale into bottles sooner rather than later to preserve the hop flavor. In this case, contrary to the common mantra, longer fermentation time is not always better. If the fermentation temperature is below 60 degrees the beer may benefit from being warmed to about 70 degrees after the initial fermentation is complete. This is called a diacetyl rest.

So where does the "two consecutive readings" rule come from?

Typical hydrometers can be read repeatedly to one gravity point. Also, one gravity point of sugar will add about 1 volume of CO_2 to the beer. Therefore two consecutive readings indicate that you are within one point of final gravity. If you carbonate your beer to 2 volumes of CO_2, the worst case would be a product carbonated to 3 volumes of CO_2 which is still a reasonable amount of carbonation. This is a very reasonable rule to follow, but the time between the readings is a variable that can be addressed to better indicate when the beer is done.

How many days should be used between readings?

There are two main variables at work here. These are the original gravity of the beer and the fermentation rate. The original gravity is easy to determine, but there are a plethora of conditions that effect fermentation rate. The goal is to determine when the beer is within one point of final gravity. The fermentation rate itself is an unknown, but we do know that it will progress roughly logarithmically. Another way to think about this is that each day a percentage of the remaining sugar will be converted to alcohol. Each day there is less sugar, meaning the fermentation rate will decreases with time.

The resolution of hydrometers and refractometers can lead to an incorrect assessment of fermentation completion. Consecutive readings a day a part do not necessarily mean that the gravity of the beer has stopped dropping.

The Answer

Crunching the numbers can lead to this same conclusion, but the following table shows how many days apart readings should be taken based on the initial gravity and the number of days from the start of fermentation.

$$Gravity = OG \times e^{-\tau t}$$

Gravity is the number of gravity units above final gravity.
OG is the original gravity (corrected for final gravity)
τ is the fermentation rate in percent per day
t is the time in days.

Readings can be taken 1 day apart until day 10 of fermentation.

Wait 2 days between readings between day 11 and 20 of fermentation.

Wait 3 days between readings between day 21 and 25 of fermentation.

Wait 4 days between readings between day 26 and 30 of fermentation.

Wait 5 days between readings if fermentation continues longer.

Calculating Fermentation Time

Most brewers will tell you that fermentation time can't be calculated but that for a known recipe and know fermentation conditions it can be approximated. However, I think with sufficient data the date of completion can be calculated and the final gravity can be calculated to within half a gravity point.

Yes, to the day, and within 1.0005 of the final gravity.

Measurements of fermentation of several beers I have brewed show that if fermentation temperature is maintained the fermentation rate is exponential. Data from three measurements and an exponential fit matches very well. The main source of error is the resolution of the hydrometer which is very accurate, but can only be read to about one gravity point. The second source of error comes from the fermentation starting rate. If the fermentation is slow to start then the original gravity and pitch time cannot accurately be used as the first point of the data.

Putting it into practice.

I brew and bottle on Friday or Saturday. To get a hoppy pale ale that I want to preserve the hop flavor gravity prediction can be used to get the beer into bottles as soon as possible. (As an alternative hop tea can be added when the beer is packaged. See "Bitter Without the Boil.") The days in between can be used for data collection. If I measure the specific gravity 24 hours from the pitch, then on the following Tuesday and Thursday there is sufficient data to perform an exponential fit.

The basic steps are as follows:

1) Record the specific gravity, date and time of three days.
2) Enter the data into excel. Add a column for the number of days from the time of pitching the yeast. Add a column for the specific gravity less the final gravity.

3) Add an exponential trend line for the new columns that you have created.

4) Adjust the final gravity until the best fit R squared value is achieved (as close to 1 as possible).

5) Add a column with the equation that has been derived and plot the subsequent days.

6) On the day that the new equation drops below 1 your beer is ready to bottle.

7) Check the gravity of the beer on that day and if you did everything right it will be on the money.

How Long Does it Take to Carbonate?

Most beers will be fully carbonated after three weeks at 70°F followed by 1 week in a refrigerator.

... but you might not have to wait that long.

One solution to beat the wait is to open a bottle of beer every few days until they are at the carbonation level of your liking. However this results in tasting many disappointingly flat beers. What we need is a way to measure the level of CO_2 in the bottle without opening more than one. As luck would have it, there is such a method!

If a balloon is placed on top of one of the bottles and secured, it will collect the gasses that are produced during this carbonation period. It's simple. When the balloon begins to deflate, the beer is carbonated.

Balloons are not great at holding pressure over a time period of two weeks. Consider how long party balloons stay floating. If you don't get them the day of, or day before the party you're going to have lackluster balloons. In the same way that helium leaves the party balloons over time, the CO_2 will leave the carbonation indicator you have set up.

This works for indicating when the yeast have completed metabolizing the priming sugar, but doesn't give you any indication of how much CO_2 was produced.

To measure the CO_2 production you will need to vent the generated CO2 into a graduated vessel for measurement.

Trub Loss

In an effort to compare an all grain beer to one brewed entirely with malt extract, two American Pils were brewed. These two beers were brewed as close to each other as possible with the exception that one was all grain, and the other was extract based. The same hop tea or hop extract was used for bittering.

Overall the beers were very similar. The most surprising difference was the trub loss.

1.5-gallon batch size
OG: 1.069
FG: 1.017
IBU: 32
SRM: 4

Fermented at 56°F

All Grain		Grain / Extract	Extract	
$6.42	3.5lbs of Pils	Grain / Extract	2.2lbs of Briess Pils DME	$7.42
$0.90	1/2 oz of cluster	Hops	1/2 oz cluster	$0.90
$6.80	1 pkg of W34/70	Yeast	1 pkg of W34/70	$6.80
$14.12		Total		$15.12
11 bottles		Yield	15 bottles	
$1.28		cost per bottle	$1.01	

Table 29 - All Grain vs Extract

Making the extract took 12 minutes from weighing the first ingredient to pitching the yeast. The all grain batch took 5 hours.

Brewing Engineering

All grain methods typically lose 20% of the beer to trub, although with this batch it was 29%. It is common to start with 6 gallons at the end of the boil, transfer 5.5 to the fermentor and have 5 gallons for bottling. When brewing with extract the trub losses are lower because the extract has already been boiled once. This means that the extract has already gone through one hot break. This batch was fairly typical at 10% trub loss.

After bottling the beer the remaining slurry was weighed. Because the two had the same starting gravity, yeast strain, and fermentation conditions the total number of yeast cells produced should be nearly identical. The all grain batch had 1,124g of slurry remaining while the extract batch measured it at only 326g. This means that the cell density of the extract batch was about 3 times higher than that of the all grain. This means when storing yeast, the same number of cells from extract beers take up about one third of the space in your fridge compared to all grain.

But how do they taste?

I asked 10 of my brewing friends to rate these two beers. None of them preferred the all grain beer, 6 preferred the extract beer, and the other 4 liked them equally well! A larger sample set would be needed to draw a real conclusion, but this makes extract brewing look very enticing.

Fermentation Control

9

Swamp Cooler

Using an ice or water bath as a swamp cooler is common practice by the home brewer to allow fermentation at temperatures lower than the ambient air temperature. Swap coolers are very effective at lowering fermentation temperature and maintaining temperature without a dedicated refrigerator. Water conducts energy much better than air. The thermal conductivity of air is 0.025 Watts per Meter degree Kelvin, while water is 0.5. In other words, water conduct heat twenty times better than air. During fermentation the yeast creates heat energy. When the fermentor is surrounded by air this heat will raise the temperature of fermentation by 4-10°F above that of the air. When the fermentation chamber is in a water bath, the water acts like a heat sink for the fermenting beer. This essentially expands the radiated area of the fermentor and air surrounding the beer to the area of the bin used for the swamp cooler. It also creates a much larger thermal mass for the yeast to heat. The large water volume therefore helps stabilize the temperature of the fermentation. This makes the temperature of the fermentor more stable, and also radiates the thermal energy produced by the yeast into the air via the large volume of water.

Bottom line: Swamp coolers provide much more stable temperature, even if you aren't looking to lower the fermentation temperature.

When using ice to lower the fermentation temperature the temperature will drop very quickly as the ice melts in the water bath. If this drop in temperature is more than a few degrees yeast will start to prepare for dormancy. Once the temperature rises they will start fermenting again, but this is at a cost of energy. Unwanted byproducts occur as the cells shift back and forth between these phases.[54] For this reason it's best not to use more than one gallon of ice in 5 gallons of water.

---

[54] http://www.fasebj.org/cgi/content/meeting_abstract/24/1_MeetingAbstracts/833.6

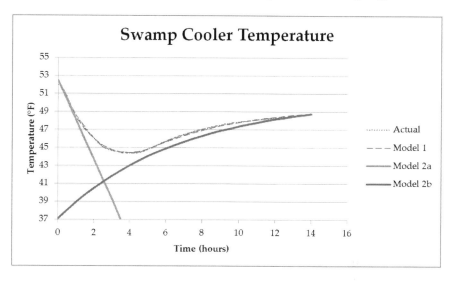

The temperature of the water in the swamp cooler goes through two distinct phases. Cooling, when the ice is added, and warming, after all the ice has melted. These phases can be modeled fairly simply using thermal conductivity. The rate of temperature changes is proportional to two factors: the thermal conductivity of the materials and the area of contact. When there is ice in the swamp cooler, at any point in time the temperature of the water is being pulled down to freezing by the ice and up to the ambient air temperature by the surroundings. This is shown in the graph above as Model 1. This model is extremely close to the actual temperature. For home brewing, this is much more accurate than necessary. A simplified model can be applied that is much easier to work with.

There are two important considerations during this phase: time and temperature. The time it takes for the ice to melt is proportional to the size of the block of ice. Ice cubes will melt in a matter of minutes. A 20oz bottle of water will take about 45 minutes, and a 1 gallon container of ice can take nearly 4 hours. Heat is first transferred by the state change of the ice from solid to liquid. This means that the water in the swamp cooler will drop in temperature as the ice changes from

Brewing Engineering

solid ice at 32°F (0°C) to water at the same temperature. This can be evaluated as heat capacity.

If you are just interested in how you can figure out what temperature your swap cooler will be at, skip ahead. If you are curious to see the derivation of the model, keep reading.

The heat of fusion of water is 334 kJ/kg.[55] The heat capacity of liquid water near freezing is 4.121 J/(g°K)[56] while the heat capacity of ice is 2.11 J/(g°K) [57] which is nearly half. The drop in temperature, discounting warming by the surroundings, is a function of temperature and mass.

MW = Mass of Water

MI = Mass of Ice

TW = temperature of Water (in °K)

273 = temperature of ice (in °K)

T = resting Temperature

$$\frac{4.121 \times TW \times MW + 2.11 \times 273 \times MI + 334 \times MI}{4.121 \times (MW + MI)} = T$$

MW = 20kg (20 liters)

MI = 3kg (3 liters when liquid)

TW = 284°K (52°F, 11°C)

273 = temperature of ice (in °K)

$$\frac{4.121 \times 284 \times 20 + 2.11 \times 273 \times 3 + 334 \times 3}{4.121 \times (20 + 3)} = T$$

$$\therefore T = 276°K \ (37.2°F, 3°C)$$

---

[55] http://en.wikipedia.org/wiki/Heat_of_fusion
[56] http://www.engineeringtoolbox.com/water-thermal-properties-d_162.html
[57] http://en.wikipedia.org/wiki/Heat_capacity

Making Great Beer Through Applied Science

The warm up phase is highly dependent on the type of container that is being used to hold the water. A cooler will hold temperature much better than a plastic bin. The ability for the container to hold temperature is based on its thermal conductivity. Typically this is measured in watts per degree Kelvin,[58] but for practical purposes it can be expressed as percentage of temperature loss per hour. This can be expressed as an exponential equation.

TA = Temperature of Air
TW = Temperature of Water
TW0 = Temperature of Water at start
$\Delta T = TA - TW0$
t = time (in hours)
TC = Thermal Conductivity (in percent change per hour)

$$TW = TA - \Delta T e^{(TC)t}$$

Determining the Thermal Conductivity of a Swamp Cooler

The temperature coefficient can be found by knowing two points on the timeline of the warming water. The process is the same whether you are working in degrees centigrade or Fahrenheit.

1. Add about 5 gallons of water, a full fermenter, and a large block of ice to the swamp cooler.
2. Once the temperature has dropped 10°F (6°C) remove the ice and record the initial temperature of the water as TW_0 and the temperature of the air as TA.
3. After six hours, record the temperature of the water, this time as TW_F.
4. Calculate ΔTW and ΔTA using the following equations:
$$\Delta T_F = TA - TW_F$$

---

[58] http://en.wikipedia.org/wiki/Thermal_conductivity

BREWING ENGINEERING

$$\Delta T = TA - TW_0$$

5. Find the Thermal Coefficient on the table below.

Delta T

DTF	5	6	7	8	9	10	11	12	13	14	15	16
1	-6.50	-7.25	-7.75	-8.25	-8.75	-9.25	-9.50	-10.00	-10.25	-10.50	-10.75	-11.00
2	-3.75	-4.50	-5.00	-5.50	-6.00	-6.50	-6.75	-7.25	-7.50	-7.75	-8.00	-8.25
3	-2.00	-2.75	-3.50	-4.00	-4.50	-4.75	-5.25	-5.50	-5.75	-6.25	-6.50	-6.75
4	-1.00	-1.50	-2.25	-2.75	-3.25	-3.75	-4.00	-4.50	-4.75	-5.00	-5.25	-5.50
5		-0.75	-1.25	-2.00	-2.25	-2.75	-3.25	-3.50	-3.75	-4.00	-4.50	-4.75
6			-0.50	-1.25	-1.50	-2.00	-2.50	-2.75	-3.00	-3.50	-3.75	-4.00
7				-0.50	-1.00	-1.50	-1.75	-2.25	-2.50	-2.75	-3.00	-3.25
8					-0.50	-1.00	-1.25	-1.50	-2.00	-2.25	-2.50	-2.75
9						-0.50	-0.75	-1.25	-1.50	-1.75	-2.00	-2.25
10							-0.50	-0.75	-1.00	-1.25	-1.50	-2.00
11								-0.25	-0.75	-1.00	-1.25	-1.50
12									-0.25	-0.50	-1.00	-1.25
13										-0.25	-0.50	-0.75
14											-0.25	-0.50
15												-0.25

Table 30 - Swamp Cooler Thermal Coefficient

The previous graph can be used as an example. At the five hour mark all of the ice had melted. The five hour point and the eleven hour point (5+6=11) can be used to find the thermal coefficient. TW_0 is 45°F, TW_F is 48°F, and TA is 52°F. Therefore ΔT_F is 3, and ΔT is 7. The coefficient is -3.5 indicating that the swamp cooler will be nearly back to ambient temperature in ⅓ to ¼ of a day. This can be seen as "Model 2b" plot on the graph.

A small cooler with a lid and filled nearly to the top with water can have a coefficient of -1, indicating it would take one day to reach the ambient air temperature.

Putting the swamp cooler to use

For a large plastic bin filled with 6 gallons of water:

- Each gallon of ice will lower the temperature about five degrees.
- Each 20 oz bottle of water will lower the temperature about one degree.
- Larger blocks of ice will maintain a more stable temperature.

Aim for a target low temperature twice what you need to account for the warm up period. (For example, if your air temperature is 70 degrees, and you want to ferment at 65 degrees, shoot for 60 degrees which would be two gallon blocks of ice.)

- Once the initial water temperature has been achieved, maintain it by adding one 20 oz bottle of ice for every degree below air temperature every 12 hours.

Aquarium Heater

A swamp cooler, or water bath, is one of the simplest and most effective ways to regulate fermentation temperature. Because of the excellent thermal conductivity properties of water the temperature of the fermentation tracks that of the water temperature within one degree. Fermentation temperature has a major impact on the flavor of the beer. Fermenting toward the high end of the yeasts range (65°F to 75°F for Ales) will add more characteristic phenols. Fermenting toward the lower end (60 to 65°F) will leave a more clean taste and can leave more residual sugar. Lagers are typically fermented even cooler at 45 to 55°F degrees, and other yeasts are designed to work best at temperatures near 80°F

Control of the fermentation temperature is therefore a wonderful tool to have in your homebrewing belt.

If you keep your swamp cooler in the basement the temperature of the water bath will likely be in the low sixties or below when no heat is added. If a way to add heat with temperature control is introduced, fermentation at temperatures above that of the ambient air are possible without any interaction during the fermentation process. The easiest, and best way that I have discovered to do this is by using an aquarium heater.

When selecting a heater for your water bath there are several factors that you will want to evaluate. The heater you select needs to be effective for heating the volume of water in your swamp cooler. A 50W heater is good for 10-15 gallon fish tank. This is a reasonable size for keeping a 5-gallon batch of beer about 10°F above air temperature, but if you want a larger temperature difference you might consider a 100W heater. The next factor is temperature. Make sure to select an aquarium heater that has a thermostatic control, not just a heat adjustment knob. During the initial phases of fermentation the yeast will produce a fair amount of heat, but toward the end it will not be producing much heat at all. In order to maintain a constant temperature it is therefore important that the heater can regulate the amount of heat it adds to the tank based on the temperature of the water. Temperature range is also an important consideration. The fermentation temperature of yeast is close to that of a typical fish tank, but check that the heater has settings for the range you will be using.

After doing a lot of research and evaluating different products, the best one I have found is the Fluval M Submersible Heater. The temperature is settable from 66 to 86 degrees which is perfect for ales. It also stretches into higher temperatures that many Saison and Belgium yeasts prefer.

I've made several beers using this technique with the Fluval M 50W Submersible Heater, including a Saison in the dead of winter. They all have come out to my liking.

Lagering Outside

To properly lager a beer the fermentation temperature needs to be near 50°F for the first two weeks of fermentation. Many home brewers have dedicated refrigerators, some with temperature controllers, or other more elaborate setups to accomplish this. A swamp cooler with bottles of ice is an alternative that others use. Without insulation ice in a swamp cooler will melt in less than an hour, and the temperature will climb back up to the ambient temperature over the course of about 12 hours. This is adequate when the goal is to ferment a few degrees below the ambient air temperature, but for lagering this might be a stretch. I don't know about you, but my home isn't kept at 55°F in the winter.

The average temperature outside, however, could be at the lagering temperatures for several months throughout the year, depending on your location. Here in Malden, Massachusetts the average temperature is between the lagering temperatures of 48°F and 58°F for about four months.

So the question is, what will happen when the swamp cooler is brought outside?

The Tou, or temperature-coefficient, of my system is 0.136°F/hour. This is the amount of energy that the system absorbs from the outside per hour. This means that 13% of the heat from the outside system is absorbed by the fermenter, and 87% of the heat in the fermenter is contained every hour. Using this Tou, the reaction of the swamp cooler can be model for a non-static temperature using an iterative process.

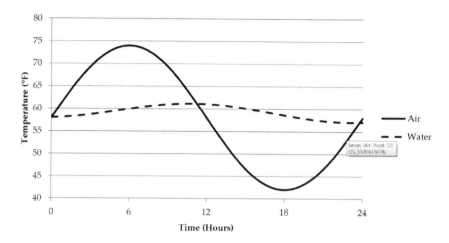

If the Tou of the fermenter can be reduced to 4% per hour then a reasonably stable temperature can be achieved. Typical camping type coolers meet this requirement when filled near capacity. The measured tou of an unmodified 22 quart cooler was 4.3% temperature loss per hour. By insolating the lid and taping it shut the performance would only improve. If you would prefer to fit your ale pail inside of a cooler then consider a 10 gallon beverage cooler.

Packaging

10

Not Just Sanitized, but Sterilized!

Short of having an autoclave in your home, this is by far the easiest way I have found to prepare bottles. It's better not to use soap to clean your bottles. The soap will leave a residue that is detrimental to the head retention of the beer. Once the bottles are rinsed out they can be placed in the oven. The racks in my oven allow over two cases of bottles to be sterilized which is perfect for a five-gallon batch of beer. Once the bottles have been laid out inside, set the oven to 350°F. After 90 minutes turn off the oven, and the next day the bottles will be cool and ready to fill with beer. Placing foil over the tops of the bottle will keep most of the air inside meaning that the bottles can even be used at a later date as long as the foil stays intact.

How It Works

Autoclaves work by a combination of heat and pressure to more rapidly kill all microorganisms, but dry heat alone can also be used to effectively sterilize glass. This process is known as Dry Heat Sterilization.[59]

---

[59] http://en.wikipedia.org/wiki/Sterilization_%28microbiology%29#Heat_sterilization

Easy Priming Sugar

Boiling and cooling priming sugar can be a pain. There are advantages to this, but how much does this extra work benefit the beer?

A further analysis of why these steps are performed might help develop a better process. I think it is possible to retain all of the benefits of boiling and cooling priming sugar without adding any time to the process.

The benefits are twofold. Boiling the water and priming sugar allows the sugar to dissolve more easily and kills bacteria and other microorganisms that may have been introduced. Cooling it keeps the yeast from being killed by the boiling liquid, and also keeps flavors from the plastic bucket from being leached out into the beer. HDPE used for food grade plastic buckets is rated for temperatures up to 190°F.[60] Exceeding that temperature could leach unwanted flavors out of the plastic and into the beer.

Killing Bacteria

Most bacteria can be killed by flash pasteurizing.[61] [62] Tap water contains very little bacteria to begin with because there are no nutrients. For bacteria to grow, both nutrients and water are required. Dry sugar also contains very little bacteria because there is no water. Therefore the amount of bacteria that needs to be killed is small. Heating to 165°F or above for a minute or longer is sufficient for most brewers.

Not Killing Yeast

---

[60] http://www.usplastic.com/catalog/item.aspx?itemid=23220&catid=752
[61] http://en.wikipedia.org/wiki/Flash_pasteurization
[62] http://www.fsis.usda.gov/factsheets/Danger_Zone/index.asp

Yeast will be killed nearly instantaneously if shocked with 165°F degree water, so the common thought is that the priming sugar needs to be cooled before adding it to the bottling bucket. While it is true that the yeast will be killed at 165°F, it's also true that the temperature drops very quickly as cold beer is added to the bucket. Yeast, like most bacteria, will thrive at 110°F. (However, it will produce off flavors if fermented for a period of time at that temperature which is why most beer is fermented at 65°F and below.) The beer will likely be about 65°F or cooler at the time of bottling. 1 half gallon of beer plus 1 quart of hot sugar water at 165°F will yield a combined temperature of 98°F.

The Process

1) Add your priming sugar and water to a microwavable container. I prefer a mason jar. (For the correct amount of water and sugar to use so as not to change the ABV of the beer see Matching ABV with Priming Sugar.)

2) Microwave for one minute with the lid off.

3) Remove from the microwave, secure the lid and swirl to dissolve most of the sugar.

4) Remove the lid and place back in the microwave for another minute.

5) Repeat steps 3 and 4 until the sugar is dissolved, and the temperature is above 165°F.

6) Start the siphon of beer into the bottling bucket.

7) Once there is approximately half a gallon of beer in the bucket add the sugar solution being careful not to splash the liquids.

Matching ABV with Priming sugar.

One thing easily neglected when priming beer is that adding dry sugar or malt extract will increase the level of alcohol of the beer. It would be easy to assume that adding a few ounces of sugar will not affect the alcohol level much considering pounds of grain or extract were used in making the initial wort.

However a few ounces of priming sugar could significantly alter the alcohol content of the beverage.

My fear was that dissolving the sugar in additional water would water down the beer. My wife prefers to drink not more than 4% ABV. With this in mind, the target for one of my brews was 3.5% ABV and about 3 volumes of CO_2. The OG and FG were right on the money (a light 1.030 and dry 1.003 respectively) leaving the ABV at 3.6%. An online priming sugar calculator indicated that 3.3oz of sugar should be added for this 3-gallon batch. 3.3 oz of sugar in 3 gallons of beer has a specific gravity of 1.003. That's 10% of the initial 1.030. The resulting ABV by adding this sugar is 4.0%!

So either assume that the priming sugar will increase the ABV by 0.5%, or add water when priming.

To maintain the ABV of the beer the sugar should be in a solution equal to that of the original gravity of the wort. Knowing that white sugar adds 46 gravity points per pound per gallon, with a little algebra the equation can be worked out.

$$\frac{oz \; of \; sugar}{original \; gravity \; in \; gravity \; points} \times 46 = cups \; of \; water$$

Gallons of Beer

OG	1	2	3	4	5	6	7	8	9	10	11	12
1.020	2.3	4.6	6.9	9.2	11.5	13.8	16.1	18.4	20.7	23.0	25.3	27.6
1.025	1.8	3.7	5.5	7.4	9.2	11.0	12.9	14.7	16.6	18.4	20.2	22.1
1.030	1.5	3.1	4.6	6.1	7.7	9.2	10.7	12.3	13.8	15.3	16.9	18.4
1.035	1.3	2.6	3.9	5.3	6.6	7.9	9.2	10.5	11.8	13.1	14.5	15.8
1.040	1.2	2.3	3.5	4.6	5.8	6.9	8.1	9.2	10.4	11.5	12.7	13.8
1.045	1.0	2.0	3.1	4.1	5.1	6.1	7.2	8.2	9.2	10.2	11.2	12.3
1.050	0.9	1.8	2.8	3.7	4.6	5.5	6.4	7.4	8.3	9.2	10.1	11.0
1.055	0.8	1.7	2.5	3.3	4.2	5.0	5.9	6.7	7.5	8.4	9.2	10.0
1.060	0.8	1.5	2.3	3.1	3.8	4.6	5.4	6.1	6.9	7.7	8.4	9.2
1.065	0.7	1.4	2.1	2.8	3.5	4.2	5.0	5.7	6.4	7.1	7.8	8.5
1.070	0.7	1.3	2.0	2.6	3.3	3.9	4.6	5.3	5.9	6.6	7.2	7.9
1.075	0.6	1.2	1.8	2.5	3.1	3.7	4.3	4.9	5.5	6.1	6.7	7.4
1.080	0.6	1.2	1.7	2.3	2.9	3.5	4.0	4.6	5.2	5.8	6.3	6.9
1.085	0.5	1.1	1.6	2.2	2.7	3.2	3.8	4.3	4.9	5.4	6.0	6.5
1.090	0.5	1.0	1.5	2.0	2.6	3.1	3.6	4.1	4.6	5.1	5.6	6.1
1.095	0.5	1.0	1.5	1.9	2.4	2.9	3.4	3.9	4.4	4.8	5.3	5.8
1.100	0.5	0.9	1.4	1.8	2.3	2.8	3.2	3.7	4.1	4.6	5.1	5.5
1.105	0.4	0.9	1.3	1.8	2.2	2.6	3.1	3.5	3.9	4.4	4.8	5.3
1.110	0.4	0.8	1.3	1.7	2.1	2.5	2.9	3.3	3.8	4.2	4.6	5.0

Water used for priming (in cups)

Sugar to achive 3 volumes of CO_2 (oz)

1	2	3	4	5	6	7	8	9	10	11	12

Table 31 - Water and Priming Sugar

Brew Lab Equipment

11

Hydrometer vs. Refractometer

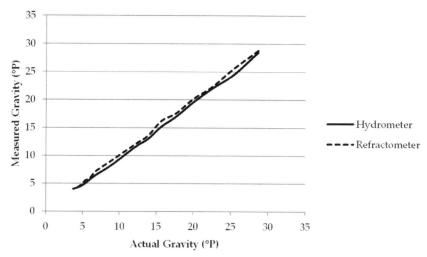

The refractometer and the hydrometer work equally well for measuring original gravity of wort. The refractometer used for this comparison has a demarcation resolution of 0.2°P while the hydrometer was marked every 0.5°P. The resolution limit of the refractometer is better, but the actual accuracy of the tools were about the same at 0.5°P

Hydrometers work by measuring the relative density of a liquid. Brewers use this density, assuming that it is a solution of sucrose and water, to determine the sugar content. Refractometers measure the refraction index; as with a hydrometer, the assumption is made that the solution is water and sucrose. Because sucrose has a very similar density and refraction index when compared to other sugars this is a reasonable assumption.

Compound	IOR	Density
Water	1.333	0.9982
Sucrose	1.561*	1.587
Ethanol	1.361	0.789
Maltose		1.54
Glucose		1.54
Isopropanol	1.3776	0.786
Methanol	1.3288	0.7918

Table 32 - IOR and Density

Value of a theoretical 100% solution by volume used for calculating the Brix value.

There are some common sources of error between the two devices. The liquid must be well homogenized before a sample is taken. Stratification is especially common when working with liquid malt extracts.

There are some sources of error unique to the refractometer. Because a very small sample of wort is used any contaminate left on the prism will add error to the measurement. Cleaning the prism after use is important so that sugar residue does not skew subsequent readings.

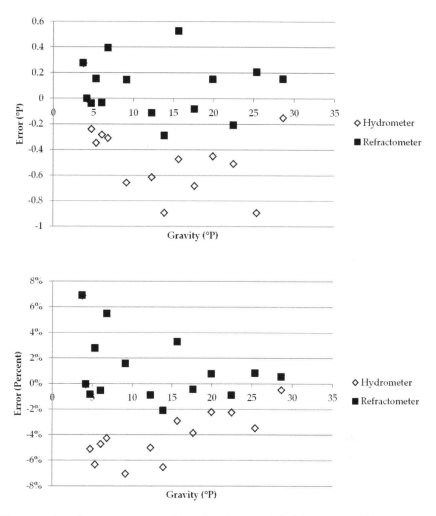

The graphs above were produced using serial dilutions. The starting volume was 100 ml of a 23° Plato wort. Each dilution was achieved by removing 10 ml of wort and replacing it with 10 ml of water.

Selecting a Microscope

The Amscope Biological Binocular Microscope is everything I thought it would be. It's not nearly as nice as my brother's professional Olympus, but for the money, the Amscope can't be beat. The microscope comes with just about everything I wanted and then some. There is oil for the immersion lens, two color disks to improve resolution when using the higher magnification settings, and a spare bulb. One thing that I was pleasantly surprised by is that not only can the width of the beam of light be adjusted with the diaphragm, but the condenser lens for the lamp can be focused. The combination of these two adjustments can create very crisp images at the dry 400x power. The 1000x oil immersion is even more clear!

A couple of things I learned the hard way about microscopes. 1) Especially at higher magnification cover slip thickness is important to image quality. 2) With oil immersion the cover slip must be very thin. If the cover slip is too thick then the lens will not be able to get close enough to the cover slip to focus. These objectives are designed to work with 0.16 mm cover slips, also known as a #1.5H

The optics are good, but not great. If you want a clear image above 1000x then you will likely need to spend several times the money on a professional scope. The 20x eyepieces work fairly well, but the resolution is only slightly better than with the same objective and the 10x eyepiece.

For counting yeast cells this scope is more than adequate. I am happy with my purchase.

Cell phone photos don't do the microscope justice. It is much more clear when looking through the lens of the scope.

In the picture below the sausage shaped cell is likely a bacteria. Luckily this was a rare find when looking at my yeast.

Making Great Beer Through Applied Science

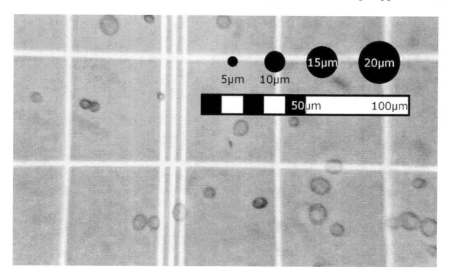

Hemocytomeres

The price of a hemocytometer ranges from $20 to over $200. The Bright-Line sells for about $260 on Amazon right now, but luckily I was able to borrow one to evaluate. For my own use I purchased a $33 hemocytometer made by Qiujing.

The difference under a microscope is night and day!

I was absolutely amazed how much the quality of the slide has to do with the crispness of the image! With the Brite-Line subtle details that appear to be birth scars on viable yeast cells were visible, and texture *inside* the nucleus of the dying cells could be seen. The Qiujing will work fine for counting viability, but the process is much slower. It's hard to find a focal depth where the cells are clear and the lines can be distinguished. Also, smaller objects, such as bacteria, are harder to resolve.

But for homebrew, I can't justify spending more on a slide than I spent on the microscope. The squares aren't exactly square, but the ruling is close enough not to add much error to the cell count. There are other sources of error that can cause greater deviation in measurements.

With some practice, and learning how to best adjust the microscope the Qiujing is more than adequate. If you are a microbiologist, or a lab technician you may be disappointed with the quality. However, for the money, I don't think you can beat it. It has been of great value to me!

Brew Lab Technique

12

Overview

While working in a brew lab many things are learned the hard way. Hopefully sharing some of my experiences in the lab will prevent you from making the same mistakes.

Be careful

The Internet has a wealth of information on procedures and techniques for a variety of measurements, but keep in mind that not all of these are correct, or even safe! Read more, experiment less. Use common sense. Talk to people who know what they are doing. Much of what is available on the Internet comes in small scraps of information, leaving a majority of the procedure left to assumptions. This can be a disaster waiting to happen. Proceed with caution.

Measure thrice cut once

Running every experiment three times is a good idea. This is especially true if the experiment will take a long time. Running the experiment just once can lead to false conclusions because there will be no way of knowing if something went wrong. If the experiments are run in duplicate, the difference between the two will indicate the amount of error that your results may have, however it doesn't indicate which result is correct. When running experiments in triplicate the outlier can be examined for the source of the error and discarded if appropriate.

Pipettes

You probably know that pipettes are calibrated to dispense an accurate volume when the bottom of the meniscus of the sample lines up with the graduated line. But, did you know that about 5% of the volume of the sample remains in the pipette clinging to the glass after the sample is dispensed? The dispensed volume will be accurate because they are graduated with this in mind. However if you are not careful this can be a problem. If you wash your pipette in alcohol, and then rinse it in sterile water you can be confident that the chances of contaminating your sample are very low, but the walls of the pipette are wet. If you simply draw the sample into the pipette your sample will be diluted by 5%. Instead of just drawing into the pipette once, draw and expel the sample twice. The error due to dilution will be much less at about one quarter of one percent.

Cleaning test tubes

To clean test tubes I typically rinse them three times with water, and then fill them half way with alcohol and give them a shake. I then pour the alcohol into a bottle used only for washing. The test tubes are then inverted and allowed to air dry. This works very well, as long as the tubes are allowed to dry completely. If there is even a thin coating of alcohol in a test tube, and a yeast sample is added, the yeast will be killed quickly. It is especially important to be mindful of this when working with samples for cell counts which can often be one milliliter or less.

Focusing a Microscope

There is more to focus than just the focus knob. First, at the lowest magnification, adjust the direction of the light. This can be accomplished by closing the diaphragm all the way and then moving the lens assembly directly below the stage. Focus the slide at this point. Next adjust the focus of the light. This is done by rotating the lens assembly directly below the stage. Turn until you see what looks like fine film grain in the center of the slide. Now you'll want to get your sample in focus. Step up the magnification adjusting the focus at the intermediate steps. Once the yeast is focused with the diaphragm all the way closed you'll notice that the edge of the cell has a halo. Open the diaphragm and adjust the focus until this disappears.

How to Make Your Own SRM Tester

SRM, or the Standard Reference Method, is a widely adopted method of measuring the color of beer. The values of color range from a 1 for a light pilsner to over 40 for a dark stout. The BJCP style book uses the SRM color to identify the color that a beer of a select style should meet, and it is also a common metric that can be used for cloning a beer. Having an accurate way to measure this can help you achieve a better product and take another step toward producing award winning beer.

SRM is ten times the log of the transmission ratio of blue light at 420 nm. The definition of SRM shows that dilution of the sample can be used to scale the SRM value.[63] Furthermore, Beer's Law (The scientist, not the beverage) shows that transmission is logarithmically related in the same way that SRM is calculated.[64]

There are methods in practice that use beers with known SRM values at different dilutions to make a set of SRM reference. That sounds like a lot of work, and a waste of beer to me.

What I propose is using a dark liquid that almost everyone has in their house, and diluting it until it matches the color of the beer. The amount of water needed to dilute the liquid can then be measured and from that, the SRM value can be determined.

So what is this mystery liquid? Soy Sauce! The color content is very close to that of beer, and only a very small amount is needed for the test. Kikoman's low sodium and regular both have the same apparent attenuation of light. Testing indicates that at a dilution of 38:1 the liquid has an SRM of 8.

Two glass 250 ml graduated cylinders are a convenient way to measure the amount of added water, and to view the beer. The graduated

---

[63] http://en.wikipedia.org/wiki/Standard_Reference_Method
[64] http://en.wikipedia.org/wiki/Beer%E2%80%93Lambert_law

cylinder also makes a useful container for specific gravity measurements.

Conveniently, the volume of the water can also be measured by weight with a gram scale. One milliliter of water weighs one gram.

SRM	soy sauce tsp	ml	water tsp	ml	SRM	soy sauce tsp	ml	water tsp	ml
1	0.25	1.2	76	375	20	1.00	4.9	15	75
2	0.25	1.2	38	187	21	1.00	4.9	14	71
3	0.25	1.2	25	125	22	1.00	4.9	14	68
4	0.25	1.2	19	94	23	1.00	4.9	13	65
5	0.25	1.2	15	75	24	1.00	4.9	13	62
6	0.25	1.2	13	62	25	1.00	4.9	12	60
7	0.25	1.2	11	54	26	1.00	4.9	12	58
8	0.25	1.2	10	47	27	1.00	4.9	11	55
9	0.25	1.2	8	42	28	1.00	4.9	11	54
10	0.25	1.2	8	37	29	1.00	4.9	10	52
11	0.25	1.2	7	34	30	1.00	4.9	10	50
12	0.25	1.2	6	31	31	1.00	4.9	10	48
13	0.25	1.2	6	29	32	1.00	4.9	10	47
14	0.25	1.2	5	27	33	1.00	4.9	9	45
15	0.25	1.2	5	25	34	1.00	4.9	9	44
16	0.25	1.2	5	23	35	1.00	4.9	9	43
17	0.25	1.2	4	22	36	1.00	4.9	8	42
18	0.25	1.2	4	21	37	1.00	4.9	8	40
19	0.25	1.2	4	20	38	1.00	4.9	8	39

Table 33 - SRM Equivelents

Harvest Yeast From a bottle

There are numerous strains available to the home brewer that cover just about the whole gamut of beers anyone could care to brew, but that doesn't stop serious brewers from adding bottle harvesting to their bag of tricks. Yeast strains that cannot be purchased can be obtained from bottle harvesting, such as from a bottle conditioned Belgium beer. You can also harvest from a fellow home brewers beer that was made with exclusive yeast. Other motivations are the cost savings, and the sheer challenge.

In order to successfully harvest yeast from a bottle conditioned beer, the yeast must be viable, and not contaminated. That seems like a no brainier, but it is important to consider. If you have a microscope it's worth checking before spending much time with the yeast. If you don't have a microscope I highly recommend that you start by plating the dregs. All of the dregs that I have looked at contained at least some bacteria or mold spores, and in many cases there have been enough to dominate the culture.

One good thing about yeast is that it just does its own thing. If you want you can check it periodically, but if you forget about it for a few days it will just keep going on its own. The only real cost is a little time and some dry malt extract. This method will take five short interactions over a couple of weeks, and 200 grams of DME.

The number of viable cells can vary quite a bit from one bottle to the next. In my home brews the cell count ranges from about 300 million to 2 billion cells suspended in each bottle. The target number of viable cells used by some brewers is 1 million per ml which is about 350 million cells,[65] although the cell count can be much lower. A beer I recently harvested from had only 7 million viable cells. When working

---

[65] Brew Like a Monk

with a bottle with very few viable cells it could take a week to see activity even when starting with a few milliliters of wort!

To keep the numbers round let's just use 1 billion total cells. The viability will likely not be great, so let's assume 50%. That's 500 million cells.

2 billion cells in 30 ml yields a pitch rate of 67 million per ml.

To make a 1.036 gravity wort, 3 grams of DME can be added to 30 ml of water. Fermentation of this wort will yield about 3.7 billion cells. Now that we have healthy yeast cells, we can make a larger step. 300 ml is conveniently up to the bottom of the neck of the 12 oz bottle. 30g of DME should be used to make the 1.036 wort. This will yield 15 billion cells, and then you could go as high as 2 liters yielding 70 billion cells (½ gallon of water with 1 cup of DME), but two steps of 750 ml and 1 liter will yield 100 billion cells.

Of course you will want to keep things as sterile as possible during this process. A foil cap on the bottle is probably a good idea. Also, once the bottle has been opened it is ripe for infection so you want to start this process the same day that the beer is poured if possible.

If you wanted to have the yeast ready as soon as possible the minimum timeline is just over a week, but if you are a busy person, you could stretch the time line out to a month without any real consequence.

Here's how to do it:

>1) Add to the beer bottle containing the yeast 3 g of DME and 30 ml of water. Let it ferment for four days. (It can take a while for yeast to become active after the long period of dormancy.)
>
>2) Fill the beer bottle to the start of the neck with water and 30g of DME. Let it ferment for two days.

3) Shake the bottle to suspend all the yeast and pour it into a 2 liter or larger container. Add 75 grams of DME and fill to 750 ml. Let it ferment for two days.

4) Add 100 g of DME and fill to 1 liter of water. Let it ferment for two days.

5) Crash, decant, and store your 100 billion cells of yeast.

Isolating Cells by Plating

Plating is a wonderfully easy way to separate the good from the bad. After only two days distinct round colonies of yeast should be visible. These are what you will want to remove with a pipette to start a larger culture.

One of my recent batches of beer was used to propagate the S-04 strain. When it came time to harvest the slurry I grabbed what I thought was the sterilized jar I had prepared, but at the end of the day I realized I had not grabbed the intended jar. The jar I had picked up from the counter was in line for the dish washer and had only been rinsed out after being used as a jam jar. I saved the slurry for a week, just in case things might work out for me, but when I did the cell count I found that the viability was very low, and there were significant signs of cocci bacteria. Later, over the course of about a week I drank some of the beers and save the dredges in a container in the fridge. Unfortunately, dredges, without the protection of the beer, spoil quickly. When I looked at the dredges under a microscope half of the cells were the characteristic finger shaped mold cells. This was not going to work for bottle conditioning directly. You would think at this point I would give up. S-04 is only about $4 at my local home brew store. But I didn't. In the spirit of experimentation and for the experience I pressed on. I diluted and plated the yeast so that I could separate good yeast colonies from the mold colonies.

Making The Plates

Isolating a single yeast cell isn't as hard as it might sound, and is very possible even without a microscope. Your first job will be to make gelatin plates. You will only need one for each strain. Petri dishes are available on the internet for about fifty cents each, but it's likely you already have something in your home that will work fine. The dish should be 4" to 6" (100 mm to 150 mm) in diameter and 1/4" to 1" tall (10 mm to 25 mm) It also needs to be able to tolerate boiling water. Small foam disposable picnic plates work fine. The container doesn't need to have a perfect seal. In fact, most Petri dishes are designed with vents to allow air flow over the culture. It does, however, need to be covered to prevent dust from falling onto the surface.

	1	2	3	4	6	8	10	12
Water (g or ml)	30	60	90	120	180	240	300	360
Water (US)	⅛ c	¼ c	½ c	½ c	¾ c	1c	1 ¼c	2 ¼c
DME (g)	3	6	9	12	18	24	30	36
DME (US)	½ tsp	¾ tsp	¾ tsp	1½ tsp	2½ tsp	1 Tbl	4 tsp	4 tsp
Agar or Gelatin (g)	2.5	5	7.5	10	15	20	25	30
Agar or Gelatin (oz)	0.09	0.18	0.26	0.35	0.53	0.71	0.88	1.06
Agar or Gelatin (US)	1/2 tsp	1 tsp	1 tsp	2 tsp	1Tbl	4 tsp	5 tsp	2 Tbl

Table 34 - Gelatin Plate Recipe

Combine all ingredients and microwave on high for 30 seconds. Stir with a sanitized utensil until everything is dissolved. Microwave the mixture on high for another 30 seconds which should bring the contents to a boil. Be careful, I gave a plate an extra 30 seconds recently, "just to be safe" and it boiled over. Pour the solution onto plates and cover with inverted plates. In a few hours your plate will have jelled, but it will be easier to work with if you can wait a day. Plate will keep in the refrigerator for quite a while, but the sooner you use the plate, the lower the risk of contamination. It's easy enough to make the plate a day ahead.

Plating

The first time I tried plating I thought the best way to do it would be to dilute the cells so that I would have individual colonies across the entire plate. However gelatin cannot absorb much liquid. If more than a few drops of liquid is used the yeast will be mixed across the plate making individual colony extraction impossible.

The easiest way to isolate single yeast strains is by streaking. The idea is that you start with a concentrated amount and as you streak it across the plate the cell concentration will diminish. The longer the streak the further the cells will be apart. It takes a soft touch. It's best to avoid digging into the media on the plate. This will bury the colonies beneath the surface. A glass stir rod, a pipette, the back of a spoon, or a spatula might be easier to use than a culture loop for the streaking process.

Commonly recommended streak pattern

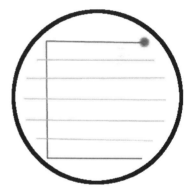

Woodland Brew streak pattern

1) Pipette one drop of your slurry or bottle dregs onto the upper right corner of the plate. (One drop is all you will need to cover the plate. The water will not be absorbed by the media so the less you use the better.)

2) Use a sterile dry pipette (or stir rod) to draw a "C" shaped line across the top of the plate and down the left hand side and across the bottom. It is important that the tool is dry. If it is wet you will end up with pools of mixed cells instead of distinct cultures. See the image above.

4) Starting from the bottom and working your way up draw lines from the left hand line across to the right working your way up the plate. These should be about a centimeter or half inch apart.

Cover the plate and put it out of the way somewhere, preferably somewhere slightly warm such as on top of the refrigerator. Try not to be curious at this point. It is very important that the plate is not moved for 48 hours. The yeast cells need to implant on the media. If they are disturbed they will continue to float around in the shallow puddle and not form individual colonies.

In 4 to 10 days you should see distinct circular colonies that are yeast. Suck these up with a pipette to grow the culture. Look out for large fuzzy growths in hazy cloud shapes. These are mold and bacteria.

Serial Dilution

Without a microscope viable cell density can be determined by serial dilution. It's a simple process, but it does require some specialized tools and about 30 minutes of work and a week of time. For this technique a known volume of slurry is diluted by factors of ten. Selections of these dilutions are plated, resulting in several plates with yeast colonies. Some plates will have too many cells to count, while others will have too few to be statistically significant. The plate with about 100 colonies on it is counted to determine the number of viable cells in the original slurry.

1) Set up one sterilized test tube for each dilution you will be making with 9 ml of sterile water.
2) Add 1 ml of slurry to the first test tube and pull in and out of the pipette 3 times to mix.
3) Remove 1 ml of liquid from the mixed tube and add it to the next tube. Mix as before.
4) Repeat step 3 until all tubes have been prepared.
5) Place 2 drops (0.1 ml) from each prepared tube of interest onto gelatin plates. (See previous section for plate preparation.)
6) Work from low density to high density. (The last prepared tube to first prepared tube.) Use a sterilized spatula to carefully spread the drop across the entire surface of the plate.
7) Allow the plates to sit undisturbed for at least 48 hours. It may take several days for the colonies to be large enough to count.
8) Count the plate that has about 100 colonies.

Depending on where the cells are taken from the cell density can be estimated to determine which dilutions to plate. Thick settled slurry contains approximately 1 billion cells per ml (1E9/ml) while a clear beer ready to bottle will have only near 1000 cells per ml (1E3/ml.) Wort at inoculation contains about 5 million cells per ml (5E6/ml) and will grow upwards of 100 million per ml (100E6/ml) at high krausen.

The table below shows the number of cells per ml based on dilution and colony count.

Number of dilutions

count	0	1	2	3	4	5	6	7
1	10	100	1.0E+3	10.0E+3	100.0E+3	1.0E+6	10.0E+6	100.0E+6
2	20	200	2.0E+3	20.0E+3	200.0E+3	2.0E+6	20.0E+6	200.0E+6
5	50	500	5.0E+3	50.0E+3	500.0E+3	5.0E+6	50.0E+6	500.0E+6
10	100	1.0E+3	10.0E+3	100.0E+3	1.0E+6	10.0E+6	100.0E+6	1.0E+9
15	150	1.5E+3	15.0E+3	150.0E+3	1.5E+6	15.0E+6	150.0E+6	1.5E+9
20	200	2.0E+3	20.0E+3	200.0E+3	2.0E+6	20.0E+6	200.0E+6	2.0E+9
30	300	3.0E+3	30.0E+3	300.0E+3	3.0E+6	30.0E+6	300.0E+6	3.0E+9
40	400	4.0E+3	40.0E+3	400.0E+3	4.0E+6	40.0E+6	400.0E+6	4.0E+9
50	500	5.0E+3	50.0E+3	500.0E+3	5.0E+6	50.0E+6	500.0E+6	5.0E+9
60	600	6.0E+3	60.0E+3	600.0E+3	6.0E+6	60.0E+6	600.0E+6	6.0E+9
70	700	7.0E+3	70.0E+3	700.0E+3	7.0E+6	70.0E+6	700.0E+6	7.0E+9
80	800	8.0E+3	80.0E+3	800.0E+3	8.0E+6	80.0E+6	800.0E+6	8.0E+9
90	900	9.0E+3	90.0E+3	900.0E+3	9.0E+6	90.0E+6	900.0E+6	9.0E+9
100	1.0E+3	10.0E+3	100.0E+3	1.0E+6	10.0E+6	100.0E+6	1.0E+9	10.0E+9
110	1.1E+3	11.0E+3	110.0E+3	1.1E+6	11.0E+6	110.0E+6	1.1E+9	11.0E+9
120	1.2E+3	12.0E+3	120.0E+3	1.2E+6	12.0E+6	120.0E+6	1.2E+9	12.0E+9
130	1.3E+3	13.0E+3	130.0E+3	1.3E+6	13.0E+6	130.0E+6	1.3E+9	13.0E+9
140	1.4E+3	14.0E+3	140.0E+3	1.4E+6	14.0E+6	140.0E+6	1.4E+9	14.0E+9
150	1.5E+3	15.0E+3	150.0E+3	1.5E+6	15.0E+6	150.0E+6	1.5E+9	15.0E+9
200	2.0E+3	20.0E+3	200.0E+3	2.0E+6	20.0E+6	200.0E+6	2.0E+9	20.0E+9

Cells per ml

Table 35 - Serial Dilutions

$$cell\ density = \frac{count \times dilution}{volume}$$

A quarter ounce package of gelatin is enough to prepare three plates. The dilutions to plate can be selected by where the cells are taken.

Location of Yeast	Approximate Density	Number of Dilutions
Thick Slurry	1 Billion per ml	5, 6, 7
Beer at high krausen	100 Million per ml	4, 5, 6
Inoculated Wort	5 Million per ml	3, 4, 5
Clear Beer	1000 per ml	0, 1, 2

Table 36 - Number of Dilutions

Refractometer Corrections for Alcohol

Refractometers are a great tool for measuring the gravity of small experimental fermentations. Without a refractometer the test worts need to be fairly large. With a one gallon fermentation vessel a large enough sample can be pulled without picking up the trub at the bottom of the fermentor. A smaller size fermentation could be trouble. With a refractometer only a few drops are needed so experimental worts can be just a few milliliters.

A hydrometer is great at measuring density, and a refractometer is great at measuring refraction index, but neither the refractometer nor the hydrometer is great at measuring alcohol in a finished beer. To use either of them assumptions must be made about the amount of alcohol produced by the sugar. The most famous of these relations is the Balling observation which is covered at length in this book.

When correcting for alcohol (like in the equation below) there is more error with a refractometer than a hydrometer, but this is not often a problem. The extra error is introduced because the scale is effectively smaller due to the alcohol compensation. With a hydrometer fermentation may go from 20°P (1.083) down to 4°P (1.016). However with a refractometer the same fermentation will start at 20°P but end at 10.5°P. The 16°P change is effectively condensed to a 9.5°P change.

A refractometer is a fine tool to indicate terminal gravity -- which is indicated by equal coincident readings -- but not absolute gravity. Being able to take a reading by pulling out the air lock and taking a small sample with a pipette is attractive. It means no more messing with the lid to take a gravity reading.

There are a couple of ways that the ABV and percent sugar equations can be derived for use with a refractometer. The first, and a common method of the online calculators is to measure with a hydrometer, and then measure with a refractometer and design an equation to fit the data. A second method is to use analytical chemistry to formulate the equation.

Detailed equations based on analytical chemistry are found in the equations section of this book.

ABV without OG

It is possible to calculate ABV without knowing the original gravity of the beer.

There are several reasons why the original gravity may be an unknown. There have been a few cases where in the rush of getting the wort into the fermentor I have forgotten to take an original gravity measurement. In the United States beverages must be labeled to 0.1% accuracy ABV, but other countries, such as Belgium, have much looser restrictions.[66] With uncertainty of the ABV level the original gravity is going to be a guess.

How the tools work

Hydrometers measure the relative density of a liquid compared to pure water. Dissolved sugar is denser than water, and alcohol is lighter than water. For original gravity measurements it is assumed that all of the dissolved material is sugar. In final gravity measurements the assumption is that the change in density is caused by sugar converting to alcohol.

Refractometers measure the index of refraction of a liquid relative to water. The assumption for the Brix scale on a refractometer is that all of the change in refraction is the result of dissolved sugar. As with a hydrometer, this is a reasonable assumption for measuring original gravity. However, although alcohol is approximately as dense as water, the index of refraction is significantly different. For this reason the Brix measurement of final gravity must be converted using a known original gravity. This is a common problem for many people the first time they take a final gravity reading with a hydrometer, but we can use this to our advantage.

---

[66] Brew Like a Monk

Brewing Engineering

Making use of the tools

Because of the distinct difference in refraction index and density between dissolved sugar and alcohol the two can be differentiated. The index of refraction of water is 1.333. Alcohol (Ethanol) has an index of refraction of 1.361.[67] Sugar water has a fairly linear index of refraction as a function of solution by volume. At a theoretical 100% solution the refraction index is 1.561.[68] [69] [70] Alcohol changes the index of refraction by only 0.028 from a zero to full concentration solution while sugar would change the index of refraction by 0.228 which is about eight times more.

The Mathematical relationships.

OGH = Original Gravity as measured with a Hydrometer (specific gravity eg. 1.036)

FGH = Final Gravity as measured with a Hydrometer (specific gravity eg. 1.005)

OGR = Original Gravity as measured with a Refractometer (Brix eg. 10)

FGR = Final Gravity as measured with a Refractometer (Brix eg. 5)

FGH = 0.00628(FGR) - 0.0025(OGR) + 1.0013 (3)

Using both a hydrometer and a refractometer the original gravity can be determined

OGR = -400(FGH) + 2.512(FGR)+400.52

And to convert back to specific gravity

OGH = 0.00432(OGR)+0.9977

---

[67] http://en.wikipedia.org/wiki/Ethanol
[68] http://hyperphysics.phy-astr.gsu.edu/hbase/Tables/indrf.html
[69] http://homepages.gac.edu/~cellab/chpts/chpt3/table3-2.html
[70] http://doclibrary.com/MSC167/PRM/ICUMSA%20Brix%20Table1933.PDF

as derived from: OGR = 231.61(OGH-0.9977) reference (4)

Putting it all together:

OGH = -1.728(FGH) + 0.01085(FGR) + 2.728

More useful would be the ABV level from final gravities as measured with a refractometer and hydrometer.

ABV=1.081*FGR-0.273*FGH-0.053

Changes of 0.001 in the hydrometer reading can change the ABV by 0.5%. So it's important that this measurement is as accurate as possible. You may want to consider a precision range hydrometer.

BREWING ENGINEERING

°B	1.000	1.002	1.004	1.006	1.008	1.010	1.012	1.014	1.016	1.018	1.020	1.025
3.0	1.033	1.029	1.026	1.022	1.019	1.015	1.012	1.008	1.005	1.001	0.998	0.989
3.2	1.035	1.031	1.028	1.024	1.021	1.017	1.014	1.011	1.007	1.004	1.000	0.992
3.4	1.037	1.033	1.030	1.027	1.023	1.020	1.016	1.013	1.009	1.006	1.002	0.994
3.6	1.039	1.036	1.032	1.029	1.025	1.022	1.018	1.015	1.011	1.008	1.005	0.996
3.8	1.041	1.038	1.034	1.031	1.027	1.024	1.020	1.017	1.014	1.010	1.007	0.998
4.0	1.043	1.040	1.036	1.033	1.030	1.026	1.023	1.019	1.016	1.012	1.009	1.000
4.2	1.046	1.042	1.039	1.035	1.032	1.028	1.025	1.021	1.018	1.014	1.011	1.002
4.4	1.048	1.044	1.041	1.037	1.034	1.030	1.027	1.024	1.020	1.017	1.013	1.005
4.6	1.050	1.046	1.043	1.040	1.036	1.033	1.029	1.026	1.022	1.019	1.015	1.007
4.8	1.052	1.049	1.045	1.042	1.038	1.035	1.031	1.028	1.024	1.021	1.018	1.009
5.0	1.054	1.051	1.047	1.044	1.040	1.037	1.034	1.030	1.027	1.023	1.020	1.011
5.2	1.056	1.053	1.050	1.046	1.043	1.039	1.036	1.032	1.029	1.025	1.022	1.013
5.4	1.059	1.055	1.052	1.048	1.045	1.041	1.038	1.034	1.031	1.027	1.024	1.015
5.6	1.061	1.057	1.054	1.050	1.047	1.043	1.040	1.037	1.033	1.030	1.026	1.018
5.8	1.063	1.059	1.056	1.053	1.049	1.046	1.042	1.039	1.035	1.032	1.028	1.020
6.0	1.065	1.062	1.058	1.055	1.051	1.048	1.044	1.041	1.037	1.034	1.031	1.022
6.2	1.067	1.064	1.060	1.057	1.053	1.050	1.047	1.043	1.040	1.036	1.033	1.024
6.4	1.069	1.066	1.063	1.059	1.056	1.052	1.049	1.045	1.042	1.038	1.035	1.026
6.6	1.072	1.068	1.065	1.061	1.058	1.054	1.051	1.047	1.044	1.041	1.037	1.028
6.8	1.074	1.070	1.067	1.063	1.060	1.057	1.053	1.050	1.046	1.043	1.039	1.031
7.0	1.076	1.072	1.069	1.066	1.062	1.059	1.055	1.052	1.048	1.045	1.041	1.033
7.2	1.078	1.075	1.071	1.068	1.064	1.061	1.057	1.054	1.050	1.047	1.044	1.035
7.4	1.080	1.077	1.073	1.070	1.066	1.063	1.060	1.056	1.053	1.049	1.046	1.037
7.6	1.082	1.079	1.076	1.072	1.069	1.065	1.062	1.058	1.055	1.051	1.048	1.039
7.8	1.085	1.081	1.078	1.074	1.071	1.067	1.064	1.060	1.057	1.054	1.050	1.041
8.0	1.087	1.083	1.080	1.076	1.073	1.070	1.066	1.063	1.059	1.056	1.052	1.044
8.2	1.089	1.086	1.082	1.079	1.075	1.072	1.068	1.065	1.061	1.058	1.054	1.046
8.4	1.091	1.088	1.084	1.081	1.077	1.074	1.070	1.067	1.063	1.060	1.057	1.048
8.6	1.093	1.090	1.086	1.083	1.079	1.076	1.073	1.069	1.066	1.062	1.059	1.050
8.8	1.095	1.092	1.089	1.085	1.082	1.078	1.075	1.071	1.068	1.064	1.061	1.052
9.0	1.098	1.094	1.091	1.087	1.084	1.080	1.077	1.073	1.070	1.067	1.063	1.054
9.2	1.100	1.096	1.093	1.089	1.086	1.083	1.079	1.076	1.072	1.069	1.065	1.057
9.4	1.102	1.099	1.095	1.092	1.088	1.085	1.081	1.078	1.074	1.071	1.067	1.059

Table 37 - OG from Final Measurements

Across the top, each column is for the final gravity as measured with a hydrometer. Down the left each row is the final gravity as measure with a refractomerer. The cells in the middle are the calculated original gravity of the wort

°B	1.000	1.002	1.004	1.006	1.008	1.010	1.012	1.014	1.016	1.018	1.020	1.025
3.0	3.2	2.6	2.1	1.6	1.0	0.5						
3.2	3.4	2.9	2.3	1.8	1.2	0.7	0.1					
3.4	3.6	3.1	2.5	2.0	1.4	0.9	0.3					
3.6	3.8	3.3	2.7	2.2	1.7	1.1	0.6					
3.8	4.1	3.5	3.0	2.4	1.9	1.3	0.8	0.2				
4.0	4.3	3.7	3.2	2.6	2.1	1.5	1.0	0.4				
4.2	4.5	3.9	3.4	2.8	2.3	1.8	1.2	0.7	0.1			
4.4	4.7	4.2	3.6	3.1	2.5	2.0	1.4	0.9	0.3			
4.6	4.9	4.4	3.8	3.3	2.7	2.2	1.6	1.1	0.6			
4.8	5.1	4.6	4.0	3.5	3.0	2.4	1.9	1.3	0.8	0.2		
5.0	5.4	4.8	4.3	3.7	3.2	2.6	2.1	1.5	1.0	0.4		
5.2	5.6	5.0	4.5	3.9	3.4	2.8	2.3	1.7	1.2	0.7	0.1	
5.4	5.8	5.2	4.7	4.1	3.6	3.1	2.5	2.0	1.4	0.9	0.3	
5.6	6.0	5.5	4.9	4.4	3.8	3.3	2.7	2.2	1.6	1.1	0.5	
5.8	6.2	5.7	5.1	4.6	4.0	3.5	2.9	2.4	1.8	1.3	0.8	
6.0	6.4	5.9	5.3	4.8	4.2	3.7	3.2	2.6	2.1	1.5	1.0	
6.2	6.6	6.1	5.6	5.0	4.5	3.9	3.4	2.8	2.3	1.7	1.2	
6.4	6.9	6.3	5.8	5.2	4.7	4.1	3.6	3.0	2.5	2.0	1.4	0.0
6.6	7.1	6.5	6.0	5.4	4.9	4.4	3.8	3.3	2.7	2.2	1.6	0.3
6.8	7.3	6.8	6.2	5.7	5.1	4.6	4.0	3.5	2.9	2.4	1.8	0.5
7.0	7.5	7.0	6.4	5.9	5.3	4.8	4.2	3.7	3.1	2.6	2.1	0.7
7.2	7.7	7.2	6.6	6.1	5.5	5.0	4.5	3.9	3.4	2.8	2.3	0.9
7.4	7.9	7.4	6.9	6.3	5.8	5.2	4.7	4.1	3.6	3.0	2.5	1.1
7.6	8.2	7.6	7.1	6.5	6.0	5.4	4.9	4.3	3.8	3.2	2.7	1.3
7.8	8.4	7.8	7.3	6.7	6.2	5.6	5.1	4.6	4.0	3.5	2.9	1.6
8.0	8.6	8.0	7.5	7.0	6.4	5.9	5.3	4.8	4.2	3.7	3.1	1.8
8.2	8.8	8.3	7.7	7.2	6.6	6.1	5.5	5.0	4.4	3.9	3.4	2.0
8.4	9.0	8.5	7.9	7.4	6.8	6.3	5.8	5.2	4.7	4.1	3.6	2.2
8.6	9.2	8.7	8.2	7.6	7.1	6.5	6.0	5.4	4.9	4.3	3.8	2.4
8.8	9.5	8.9	8.4	7.8	7.3	6.7	6.2	5.6	5.1	4.5	4.0	2.6
9.0	9.7	9.1	8.6	8.0	7.5	6.9	6.4	5.9	5.3	4.8	4.2	2.9
9.2	9.9	9.3	8.8	8.3	7.7	7.2	6.6	6.1	5.5	5.0	4.4	3.1
9.4	10.1	9.6	9.0	8.5	7.9	7.4	6.8	6.3	5.7	5.2	4.6	3.3

Table 38 - ABV Without Original Measurments

Across the top, each column is for the final gravity as measured with a hydrometer. Down the left each row is the final gravity as measure with a refractomerer. The cells in the middle are the calculated ABV of the beer.

How Accurate is a Cell Count

Cell counts aren't perfect, but it will get you closer than any other method I know for estimating the number of cells in slurry...and I've heard a lot of different methods.

To perform a cell count a sample of the slurry is placed on a specialized microscope called a hemocytometer. This slide is divided into 25 "boxes" that are of a specific volume. Typically five boxes are counted and from that the average viability and cell densities can be calculated. The standard deviation, or accuracy, can also be calculated. A typical standard deviation for viability is 1%-10%. If the deviation is higher, the sample should be recounted.

Cell counts use an extremely small sample. Less than one tenth of a milliliter is placed on the slide. The actual volume of cells that are counted is typically 20 nanoliters. That's 0.00002 milliliters. On top of that the sample typically needs to be diluted twenty to forty times from the slurry density. Another way to think about it is the number of actual cells that are counted. In a typical liter of slurry there may be one trillion cells, of those, only about 500 are actually observed.

Don't let that scare you. Even with all of those factors working against you the accuracy is still very high. The good thing is that as long as the slurry is well homogenized with adequate stirring or shaking even this small sample is representative of the entire slurry.

Because viability is a ratio of cells, it is not important that the volume of cells is correct, or that the exact dilution is known. For the cell density, however, volume and dilution are important. Errors in cell density can be seen from day to day measurements of the same slurry. When I first started doing cell counts I saw a variation of about 25% in cell density. This is much better than the factor of 10 that is common from one slurry to the next, but a substantial number when trying to work with absolutes. It is for this reason that consistency is vitally important.

Flocculation can also make determining cell density difficult. When the cells are clumped up together it is hard to see exactly how may cells are present. Currently I am using both an acidic acid MB solution and a glycine MB solution to separate the cells for counting. Acidic acid works better than glycine for separating the cells however, since because it is an acid it will pull some of the methylene blue out of the cells making viability harder to determine. In most situations the glycine MB solution will separate the cells and allow for a clear cell count.

There are other variables in cell count accuracy that may be considered. These include:

- dilution error
- poorly illuminated cells due to insufficient staining.
- viable cells taking in stain due to the toxicity of methylene blue.
- non-viable cells failing to take in stain due to the cell rupturing.

In conclusion, the small error in viability is much better than the factor of ten variations I have seen in fresh slurries, and the 25% variation in cell density is much better than the factor of 10 that can be attributed to protein concentration.

Counting Yeast Cells to Assess Viability for Brewing

Pitch rates, as you have likely heard, are very important to creating the desired flavor profile in a beer. Pitching too much yeast will result in a bland lifeless flavor, while under pitching the yeast will result in hot alcohols and possible a stuck fermentation. Generally being within a factor of two is sufficient.

How to count yeast cells with a hemocytometer.

What you need:

A microscope with 400x magnification
A hemocytometer
2 test tubes or small vials
A graduated 2 ml volumetric pipette
A 250 ml (1 cup) or larger container with a tight lid.
Methylene Blue (which can be purchased at a pet store)

Step one, making a stock 0.01% Methylene Blue solution.

Don't worry, you don't need to do this often. Staining is used to distinguish the live cells from the dead cells. Methylene Blue is attracted to acids such as the deoxyribonucleic acid (DNA) in the yeast cell. Yeast cells have the ability to reject methylene blue that has entered through the cell membrane. The result is that the dead yeast cells will stain blue, while the live yeast cells will stay clear. Because methylene blue is toxic to yeast the concentration should be kept lower than 0.1%, and once the cells have been stained they should be counted within half an hour. When working with methylene blue keep in mind that it is a very dark dye that stains very well. If you spill even a drop you may have a blue dot on your counter forever.

Methylene Blue sold to treat aquariums is typically 2.303%. To dilute this to the 0.01% concentration that is recommended by White Labs.

1) add 1ml of the 2.303% Methylene blue to the 250 ml container.

2) Fill the container with 230 ml of water (230 grams, 7.77 floz, 47 tsp, or 1 cup less 1 teaspoon).

Only 0.5 ml are needed per measurement, so this container of stock solution that you have prepared will be enough for hundreds of tests.

Step two, preparing the sample.

With the hemocytometer you will only be looking at a tiny fraction of the yeast that you are evaluating, so it is important the slurry is completely homogeneous before taking a sample.

1) Shake the container of yeast until everything is completely mixed up.
2) Shake the container some more.
3) Using a pipette, remove 1ml of yeast from the slurry container. If the slurry has large particles it may block the pipette. If this happens you can use a drinking straw with the pipette pump to remove the sample.
4) Add the 1 ml sample to one of the test tubes. The accuracy of this volume measurement will have a direct impact on the accuracy of your data.
5) Add 19 ml of water to the yeast to dilute the sample 20:1.
6) Mix the sample by shaking vigorously.
7) Pipette 0.5 ml of yeast from the sample test tube to the second test tube.
8) Sterilize your pipette.
9) Pipette 0.5 ml of 0.01% methylene blue into the second test tube. Note that this dilutes an additional 2:1.

Step three, loading the hemocytometer.

1) Place the hemocytometer on a paper towel and put the cover slip on.
2) Mix the stained yeast with the pipette.

3) Remove a small amount with the pipette (you will only need about 0.1 ml, or a fraction of a drop).

4) Bring the pipette up to the hemocytometer without letting the tip touch.

5) Dispense a hanging drop and touch the drop to the hemocytometer sample loading point. Only the counting chamber needs to be filled, not the spill trough that forms an H.

Step four, counting the cells.

If you haven't use a microscope before you might want to spend some time fiddling with the controls and doing some reading about how to adjust everything properly. I was surprised by the number of adjustments on a modern microscope, and how crisp the image can be when you have everything adjusted correctly. There is a lot more to focus than just the focus knob! When you look at the cells you should note the amount of clumping. Clusters, or clumps, of cells indicate that the sample was not adequately mixed, and is thus an indication that your count may not be representative of the entire slurry. When focusing on the cells, tighten the diaphragm all the way. If the focus is too high you will see haloes around the cells. If the focus is too low the cells will look blurry. Once the cells are in focus open up the diaphragm until the blue cells are clearly blue and the clear cells still have defined membranes. If the diaphragm is opened too much, the cells will appear washed out and you may miss some in your count. Focus is important. It the focus is too high, the halo can make a dead cell appear alive.

1) Locate the upper left counting chamber. (This is called box 1.)

2) Count and record the viable cells (clear) and the dead cells (blue).

3) Repeat this process with the upper right, middle, lower left, and lower right chambers

4) If you want to have an idea of your accuracy you can perform statistical analysis on these five data points.

Step five, calculating.

1) Viability is the percentage of living cells.
v = total number of Viable cells counted adding all five boxes
d = total number of Dead cells counted adding all five boxes
viability = v / (d+v)
example: for v = 432 and d = 10
viability = 432 / (432+10) = 98%

2) viable cell density is the number of living cells per volume
df = dilution factor = 20
sf = stain factor = 2
vol = volume of boxes counted = 4nl x 5 boxes = 20nl
cd = cell density in billions of cells per liter
cd = v * df * sf / vol = v * 20 * 2 / 20
cd = v * 2
example 432 * 2 = 864 billion cells per liter

3) Pitching volume is the amount of slurry needed to achieve your pitch rate
pv = pitching volume in ml
c = billions of cells to pitch
pv = c / cd * 1000
example: c = 200 billion cells needed to pitch
cd = 864 billion cells per liter
pv = 200 / 864 * 1000 = 231 ml (or roughly 231 grams)

There is a plethora of information about counting cells, but these are some of the best suited for this task that I have found:

A wonderful tutorial with pictures of what you should see while counting:

http://www.whitelabs.com/beer/CellCounting&Viability.pdf

Step by step procedure from White Labs:

http://www.whitelabs.com/beer/cell_count.html

Some more nice photos of what to expect and some bacteria that you don't want to see:

http://www.whitelabs.com/beer/microscope.html

Making Great Beer Through Applied Science

Trouble Shooting

13

Top Ten Reasons Why Your Final Gravity is Stuck

1) Using a refractometer to measure final gravity.
Refractometers seem to be the flashy new toy that every home brewer wants to have. They allow fast measurements of the wort or beer with a sacrifice of only a few drops of beer. However, this tool, if not used correctly, can indicate an inaccurately high final gravity. Alcohol will skew the readings. The refraction index of alcohol is different than that of water, so it must be compensated for when taking a measurement. There are several calculators available online that can compensate for this for you.

2) Using extract
Not all malt extracts are created equal. Studies have shown that some extracts, especially old liquid extracts, are as little as 55% fermentable. Ray Daniels provides a wonderful description of this in his book "Designing Great Beers"

3) Not waiting long enough between consecutive FG readings
One day apart isn't always long enough. Beer that has taken longer than 10 days needs to have consecutive readings 2 days apart to ensure final gravity is reached. See the Fermentation section for more information.

4) Using a hydrometer that is inaccurate
It's very easy to knock a hydrometer out of calibration. I used to be in the habit of just dropping it back in the tube. Eventually I started getting some readings that didn't make sense. I check the hydrometer in water and it was reading 1.005! Lightly dropping the hydrometer can move the paper inside and change the adjustment.

5) Mashing at a high temperature
Each degree above 152 causes 2% less attenuation. See Mashing section for more information.

6) Using an inaccurate thermometer to measure mash temperature

If your thermometer reads 150.0 it doesn't mean it's accurate, in fact it could be off by 5 degrees putting you at 155. Also, if it is new, or if it cost a lot of money, or if it is analog that doesn't mean it's accurate. Check it in boiling water to make sure it will be close at mash temperatures.

7) Mash time too short.

Even after conversion from starch to sugar is complete the beta enzymes will continue to break down the long sugar chains into shorter ones. Mashing at 150 for 60 minutes will produce a less fermentable mash than mashing at the same temperature for 90 minutes. To control your wort's fermentability you will want to add a mash out step at 165-170 to denature the enzymes.

8) Over use of Campden tablets

If you have highly chlorinated water you might be using Campden tablets to rid your water of free chlorine. Campden may be good at removing chlorine, but it also acts as a yeast inhibitor. In fact, it is commonly used in wine and cider making to stop fermentation. Using 1 tablet per 20 gallons will reduce the chlorine and not impede the yeast much.

9) Under pitching yeast, not oxygenating well and not adding yeast nutrients

A combination of two of the three will cause the yeast to run out of nutrients and may lead to a stalled fermentation. Yeast require specific proteins to break the longer sugar chains into short chains that they can convert to alcohol.

10) Fermentation temperature

Colder fermentation temperatures are more likely to lead to a stalled fermentation, although in my experience, this is rare. Rousing the yeast by swirling the fermentor, warming it to 70°F and adding corn sugar and yeast nutrients can get it going again.

Top Ten ways to Restart Fermentation

1) Give the fermentor a swirl.
Try the easy things first. You might be able to squeeze a few more points out of the fermentation by gently coaxing the yeast back into suspension.

2) Move the fermentor to a warmer area.
Fermentation temperature can change attenuation by about 2%. That might be enough to get over the line from cloying to malty.

3) Repitch with a higher attenuating yeast
Champagne yeasts, such as EC-1118, will ferment simple sugars to completely dry. Safale S-04 is also a high attenuating yeast. At this point in the fermentation the flavors have already been added by the beer yeast, so adding this second yeast will not impact the flavor much. If the fermentation has stopped significantly short of what was expected it could be caused by a combination of under pitching and under aeration. In this case mix the yeast with one quart of water and aerate well by shaking the container until there is a think foam on top of the slurry.

4) Add simple syrup.
Sometimes the yeast needs a little kick in the pants to get going. If you are adding yeast, then it's easy to add a little extra sugar to make sure the yeast starts up.

5) Add yeast nutrients.
Especially if the beer was under pitched the yeast can run out of nutrients. It takes special proteins for yeast to convert long sugar chains found in malt extract and worts generated from high mash temperatures. The yeast nutrients will give the yeast the proteins that they have depleted.

6) Add beano

Beano from your pharmacy, or amylase enzyme from your homebrew store, will break the longer sugar chains into shorter ones. So if the yeast in the fermentor cannot digest the long chains this will help them continue their job of conversion. The only problem I have heard with this is that it works too well. You may end up at 1.000. Half a teaspoon is a good starting point for a 5-gallon batch.

Other ways to fix the high final gravity without restarting fermentation

7) Dilute the beer

A final gravity of 1.020 will taste pretty sweet, but if diluted to 1.015 it might not be so bad. The hop bitterness and the flavor will also be diluted, making it a different beer, but this may make it drinkable.

8) Add hops

A little bit of bitter will balance out the maltiness. You could add a couple of ounces of hops right into the fermentor to dry hop the beer, or make a hop tea. Either boil the hops in a approximately a 1.020 wort for 30-60 minutes to get some bitterness, or steep them for about 10 minutes in water to just get the earthy flavor.

9) Add fruit

If it's already sweet, then run with it. Fruit, by itself, is sour in beer because most of the simple sugars that they contain are fermented into alcohol, so having some malty sweetness will make a beer that the lady folk will love.

10) Bottle it.

If all else fails, or you decide to leave it for other reasons, change the style name. Maybe the recipe was for a Porter but now you have a Dunkel. Maybe your American Lager is a Bohemian Pilsener.

Adjusting Flavor

Over the course of a beer's life the flavor and aroma will change drastically. If you are anything like me the first time tasting unfermented wort was a surprise: Raw beer tastes like candy?

Because this transformation of flavor does not end until consumption it is difficult to fine-tune the final product at the earlier stages. Unlike in cooking tasting during the preparation does not directly indicate what the final product will taste like.

Pre-fermentation the sugars mask most all other flavors in the beer. Post-fermentation tasting provides a better idea of what the product will be. There are some things to keep in mind when tasting post-fermenting beer to best assess adjustments.

1) Carbon dioxide is somewhat bitter and will mask maltiness.
2) Hop bitterness and flavor will diminish with time.

Keeping these in mind there are some simple ways to adjust the product before packaging.

Hops can be added after fermentation as dry hops, to leach out aroma and flavor without bitterness. This can take weeks to achieve full potency. A faster, and perhaps more controlled, alternative to dry hopping is making a hop tea. Just as hops are boiled in the wort for varying length of time, teas can be made by boiling hops and water. A sixty minute boil will impart only the bitterness while a 15 minute boil will capture most of the flavor without the grassiness associated with dry hops. There are other options as well. Hop aroma can also be extracted using alcohol, and it is also common to use a French coffee press with 170°F (75°C) water.

It's not too late to add specialty grains after fermentation has completed. Perhaps the temperature got a little out of control with a weizen beer and the banana flavor is too pronounced. The roasted flavor of chocolate malt may be appropriate. For a five-gallon batch of beer one half pound (one quarter kilogram) of specialty grain can be steeped in about half a gallon of water. Bring the temperature to at least 170°F (75°C) to get the most of the flavor. As with all steps in the beer making process, sanitation is important.

Efficient Brewing

14

Great Beer with Little Time

Typically, an all grain brew day is five hours, at least for me. To edge out every bit of quality in beer it can take that commitment. However, not everyone has that kind of time to invest in brewing. Life changes could take you from having several hours to spend on beer to less than one hour per batch. Personally my life has changed recently with the birth of my daughter, and I've been looking at ways to make great beer on short time.

All grains brewers, including myself, often don't think highly of extract brews. Maybe it's because it seems more expensive, lower in quality, less controlled, or perhaps it just seems like cheating.

The cost of extract beer can be lower than all grain without sacrificing quality. The cost per gravity point is quite close. This can be evaluated by looking at the cost per gravity point of the two different methods. From the local home brew store dry malt extract is $4.28 per pound. This will add about 43 gravity points per pound per gallon or 43 ppg. The cost of each gravity point is $0.10. Extract can be used to replace base malt, crystal malt, roasted malts and even some specialty malts such as Pilsner and Munich. In contrast these malts, including a 75% efficiency, costs $0.08 per gravity point.

Another important consideration when evaluating cost is trub loss. When extract is produced there is a significant hot break reducing some of the protein before it gets to your boil kettle. With all grain brewing you will have trub loss to both the hot and cold break. A typical all-grain brew will lose 20% to the trub while extract will only lose 10%. An extract batch would lose one bottle of beer to trub loss for every ten while an all-grain brew would lose two per ten.

A final consideration for cost is that of energy to boil the wort. For a typical 5-gallon batch of all grain beer it is common to start with 7 gallons of water. This accounts for the boil off and trub loss to result in

5 gallons of beer to bottle. The energy required to bring this water from room temperature to boiling is nearly 7 Mega Joules of energy. The extract technique described later in this book requires less than 1 Mega Joule. Using propane, the cost per bottle is about $0.06 more for all grain than for extract.

Factoring these together an extract 12oz bottle of 5% ABV beer will cost $0.69. An equivalent all grain beer would cost $0.75. It's a small difference, but extract does come out ahead in cost. When you think about all the time put into all grain brewing, even if you love brewing is often hard to justify the additional time commitment.

The quality of extract brews is often more of a result of the brewing process rather than the actual quality of the malt extract. Novice brewers typically start with extract brews. Common mistakes made by novice brewers are often overcome after the brewer switches to all-grain. Their extract brews may have been inferior leading them to believe that the all grain process makes better beer. Novice brewing mistakes include lack of consideration for fermentation temperature and use of tap water or bottled spring water. Malt extract is made the same way an all grain brewer would make wort. The difference is instead of cooling the wort and pitching yeast, the manufacture condenses it down to syrup, or into a fine powder. This reduction is accomplished using a low pressure boil.[71] The negative pressure sucks the moisture away. DMS is produced when wort is above 140°F (60°C).[72] The low pressure boil removes the DMS that is produced better than a brew kettle.[73]

Extract brewing can give even more control than all grain brewing. It may seem that the all grain brewer has advantageous control of the

---

[71] http://www.briess.com/food/Processes/nstep.php
[72] http://www.homebrewtalk.com/wiki/index.php/DMS
[73] http://beersmith.com/blog/2012/04/10/dimethyl-sulfides-dms-in-home-brewed-beer/

wort, but that control also adds to unpredictability. In all grain brewing variations in mash temperature drive fermentability of the wort. However, uncontrolled variations in temperature will cause changes in the fermentability. Extracts are carefully made to produce a consistent product. The two largest manufacturers of malts, Briess and Muntons, both have a 75% fermentable product. Often the brewer may want higher attenuation. Simple sugar can be added to achieve this with similar results to an all grain mash at a lower temperature. Chemically, the highly fermentable glucose produced by the enzymes in a mash is identical to that of corn sugar.[74][75] Table sugar, or sucrose, is the combination of glucose and fructose which are predominate single-chain sugars in all barley wort.[76] The difference between Maltose and Sucrose is marginal. Maltose is two linked glucose molecules while the Sucrose is one glucose and one fructose.

The varieties of extracts available give the brewer enough control to make any style of beer. Extracts often have a fair amount of Carapils added for head retention. The table below lists the all grain equivalents for most common extracts.

---

[74] See section on "Wort Sugar"
[75] http://en.wikipedia.org/wiki/Glucose
[76] http://en.wikipedia.org/wiki/Sucrose

Weight	Extract	Weight	All Grain
1	Pilsner DME	1.5	Pilsner
1	Light DME	1.5	2 Row
1	Amber DME	0.1	Crystal L60
		1.4	2 Row
1	Dark DME	0.1	Chocolate
		1.4	2 Row
1	Wheat DME	1	Wheat
		0.5	2 Row
1	Munich LME	0.7	Munich
		0.7	2 Row
1	Rye LME	0.1	Rye
		1.2	2 Row
1	Maris Otter LME	0.8	Maris Otter
		0.5	2 Row

Table 39 - All Grain to Extract Conversion

Just like with all grain brewing minerals are needed in the mash to aid conversion. These minerals are condensed along with the malt sugars meaning the extract will have the mineral content of the region in which it was produced. Briess is manufactured in Chilton Wisconsin.[77] The water in this area emphasizes the bright hop flavors due to the higher sulfate content.[78] Muntons is manufactured in Bridlington and Stowmarket UK.[79] These are located north of London and east of Burton Trent, two very prominent brewing regions. The water in these regions emphasizes the malt with the higher chloride and sodium levels. Coopers is manufactured in Australia where there is a much lower mineral content in the water. For malt forward beers Muntons is a good choice and for hop forward beers Briess will serve well. For a light bodied beer Coopers is a safe bet.

---

[77] http://www.brewingwithbriess.com/Malting101/Benefits_of_Malt_Extracts.htm
[78] http://www.cityofmadison.com/water/documents/waterQuality/Well7QualityReport.pdf
[79] http://www.muntons.com/about/locations/

	Briess	Muntons
Ca	80	50
Cl	30	60
Mg	50	20
Na	10	100
SO4	60	80
Hardness	100	150

Table 40 - Extract Mineral Concentration for a 1.060 Wort

If you use water that has minerals in it the resulting beer will contain twice the mineral content that it likely needs. For extract brewing it is best to use distilled or reverse osmosis water.

Adjusting the water profile can be simple with extract beer when made with ion free water. The table below indicates the additions for several styles. These are calculated for a 1.060 (15°P) wort of Briess malt, but will be fairly accurate for a wort from 1.030 (7°P) to 1.090 (22°P) starting gravity.

Profile	SO_4^{-2}	Cl^-	HCO_3^-	teaspoons			grams		
				$CaSO_4$	$CaCl$	$CaCO_3$	$CaSO_4$	$CaCl$	$CaCO_3$
Briess	63	28	90						
Quite Bitter	181	28	90	1			4.0		
Extra Bitter	122	28	90	1/2			2.0		
Slightly Bitter	63	28	90						
Balanced	63	63	90		1/2			1.7	
Malty	63	92	90		3/4			2.6	
Very Malty	63	120	90		1			3.4	
Dublin	55	19	200			3			5.4
London	77	60	156		1/4	2		0.9	3.6
Munich	10	2	200			4			7.2

Table 41 - Salt Adjustments for Style

Dry vs. Liquid Extracts

The two products will make slightly different beers. From experience the brewer will gain a greater appreciation for the two, though for the most part they are indistinguishable in the final product. There are a couple of things to understand from a practical stand point.

Dry malt extract is easier to measure and to store partially used, that is unless you are working in a humid environment. Dry malt is extremely hydroscopic and will pull moisture right out of the air making a sticky mess. In New England in the winter dry malt works wonderfully, however in the summer liquid malt is easier to handle. Dry malt also stores very well over long periods of time.

Liquid malt can be difficult to remove from the container. If some of the water that will be used for the wort is added to the container it can aid in its removal. Also, heating the cans in a bowl of warm water can aid in the process. If the entire can is not used it will spoil fairly quickly unless stored properly. When both sugar and water are combined it makes a breeding ground for a number of bacteria. Canning is one option to reduce the chance of unwanted bacterial growth.

Bitter Without the Boil

If it wasn't for the boil, a brew day could be reduced to a 15 minute process. Even with extract brewing a typical brew day is close to two hours primarily driven by heating, boiling and cooling the wort.

For extract brewing, there are just two main reasons to boil: hop utilization and sanitation. A third that some may consider is removal of protein. However, extract has already been boiled and cooled when it was manufactured. The protein that remains is just the right amount for head retention of the beer.

In the middle ages, when beer was invented, sterilizing the wort by boiling it was a necessity, but in today's world it's overkill. Literally. When using bottled distilled water[80] there is no need to boil it. The bacteria content is minimal. Malt extract is also essentially void of bacteria. It's a food grade product that has been packaged to last years. Sugar, in high concentrations, actually acts as a preservative. Dry malt extract has no water in which the bacteria live. Liquid malt extract has a high specific gravity that applies an enormous amount of osmotic pressure on bacteria preventing them from thriving. This osmotic pressure will also prevent spores from hydrating properly.[81]

To extract hop bitterness the alpha acids must be isomerized. Historically this has been done by boiling with the wort, however boiling the hops with the wort presents a couple of problems for those trying to make great beer with little time. First, protein in the wort makes it prone to boil over, which means the brewer will need to tend the pot often. Anyone who has boiled wort is likely all too familiar with this problem. Bubbles of water vapor are trapped by the protein making foam which rises to the surface. Second, once the hops are

---

[80] See "Great Beer with Little Time"
[81] See "Rehydrating Yeast" for a parallel concept.

boiled in the wort the bitterness is mixed throughout the entire wort. Because the bitterness cannot be easily separated the boil must be conducted for every batch of beer when using traditional methods.

The solution is to split the two into shorter processes. A smaller volume of water will boil much more quickly, and if the amount of coagulant material is limited the risk of boil over is greatly reduced. Hop tea for several batches of beer can be made at one time. The tea will be used as a bittering agent. As with all bittering hops nearly all of the aroma and flavor are boiled off, making this tea useful in any beer that requires bittering. Furthermore, the type of hop used for bittering has very little effect on the final product. The cost of the hop tea can be greatly reduced by selecting inexpensive yet effective hops.

One quart of water boiled for an hour or more with one ounce of 15% Alpha Acid hops can be used to produce a bittering tea. Water may need to be added to make up for boil off. Two teaspoons of this tea will add one IBU to a 5-gallon batch of beer (10ml of tea to 20 liters of beer). A portion of the hop tea can be set aside for the current brew, and the remainder can be poured into canning jars and stored for the next brew. Personally, I like to split this between four 8 oz canning jars. Each jar holds enough bittering tea to add about 20 IBUs to a 5-gallon batch which is well suited for most styles. Because the bitterness will diminish over the course of time I typically make enough for six months of brewing.

AA	ml	tsp	cups	AA	ml	tsp	cups
5	315	64	1.33	12	757	153.6	3.20
6	379	76.8	1.60	13	820	166.4	3.47
7	442	89.6	1.87	14	883	179.2	3.73
8	505	102.4	2.13	15	946	192	4.00
9	1009	204.8	4.27	16	1009	204.8	4.27
10	631	128	2.67	17	1073	217.6	4.53
11	694	140.8	2.93	18	1136	230.4	4.80

Table 42 - Bittering Hop Tea

Dry pellet hops are also of low risk to contaminate beer. The hops themselves act as a preservative by creating an environment toxic to many microorganisms.[82] The pelletizing process creates a large amount of heat which destroys bacteria.[83] Adding hops directly from the foil packaging to the beer is low risk to contamination.

Hop flavor can be added as dry hopping. A one ounce charge will add the characteristic flavor of nearly any hop to a five-gallon batch. The amount of bittering of one ounce of dry hops is about the same as the Alpha Acid percentage when used in a five-gallon batch.

---

[82] http://en.wikipedia.org/wiki/Food_preservation
[83] http://djcoregon.com/news/2010/03/05/how-they-make-hop-pellets/

Easy Pitching

Dry yeast contains approximately 150 billion cells per package. Instead of scaling the volume of the wort, or growing the cells, extract brewing gives you a third option that is even easier. A portion of the extract can be added to create the correct pitch rate. After two days the remainder of the sugar can be added.

$$0.75 \times volume\ (liters) \times gravity\ (°P) = billions\ of\ cells\ required$$

Plato is a measure of percent sugar by weight and one liter of water weighs 1 kg. Liters times degrees Plato is therefore the same as 10 times the weight of the sugar. Making this substitution the number of grams of sugar can be solved for with 150 billion cells.

$$0.75 \times 10 \times grams\ of\ sugar = 150$$

$$grams\ of\ sugar = 2000$$

2000 grams, or 2kg, is about 4.4 lbs.

For ales 4.4 lbs (2 kg) of the extract should be added at inoculation, and for lagers 2.2 lbs (1 kg) should be added.

Name	Description	cost	Flocc	Low Temp	High Temp	Low Temp	High Temp
T-58	Belgian Ale (spicy)	low	High	59°F	75°F	15°C	24°C
S-33	Cloudy	low	Medium	59°F	75°F	15°C	24°C
S-04	British Ale (mild mineral)	med	High	59°F	70°F	15°C	21°C
WB-06	Bavarian (clove & banana)	med	Low-Med	59°F	75°F	15°C	24°C
US-05	Classic American Ale	med	Low-Med	60°F	72°F	16°C	22°C
Nottingham	Neutral	med	High	57°F	70°F	14°C	21°C
Windsor	English (Fruity)	med	Low	64°F	70°F	18°C	21°C
W-34/70	Weihenstephan (crisp)	high	High	48°F	59°F	9°C	15°C
S-23	Saflager (Malty Fruity)	high	Medium	46°F	50°F	8°C	10°C

Table 43 - Dry Yeast Comparison

Fast Brew Method

With this method you can brew beer with less than fifteen minutes of hands on time. That means while my daughter is napping I can get a batch of beer into the fermentor and mow the lawn!

Correct pitching and fermentation are vital to the success of your beer. Traditionally a large volume of wort is boiled and cooled. The cooling process alone can take hours, and if you don't have the right equipment, it can take half a day!

For extract brewing, as we have seen, the boil is entirely unnecessary.[84]

Pitching temperatures can be achieved by simply storing the distilled water in the same location that the beer will be fermented.

With this method only part of the extract is boiled. The remainder is added dry to the fermentation vessel.

The first addition of malt extract can be mixed in by pouring the wort between two buckets. This will also serve to sufficiently aerate the wort.

The second addition of malt extract can be messy if some precautions are not taken. During fermentation yeast produces carbon dioxide which is dissolved in the beer. Adding sugar will force the gas out of solution which can easily foam over the edge of the fermenter. The fermentor can be agitated to release some carbon dioxide before adding the malt. Malt should also be added about one pound at a time until the foaming has diminished. Another option to prevent the carbon dioxide from being released from the solution is to dissolve the malt extract in water before adding it to the fermentor.

---

[84] See "Bittering without the Boil"

Adding extract without boiling it first may raise some concern of infection, but the chances are very low. It takes both sugar and water for bacteriological growth and DME only has one of the two. DME is also extremely hydroscopic, so if there is any significant amount of water it would be sticky. DME is made by spraying near boiling wort into a low pressure chamber.[85][86] Apart from making an extremely fine powder with low moisture content this also ensures a sterile product. The high heat ensures that the wort is sterile, and the low pressure ensures that the chance of additional contamination is very low. Because of the way DME is made the chances of contamination are extremely low if it is still in the manufacturer's packaging. In addition, sugar acts as a natural preservative.[87][88] The high concentrations of sugar in malt extract actually make it more difficult for bacteria to grow in than a lower concentration of sugar in wort.

1) Rehydrate the yeast in the ale-pail.[89]
2) Pour the first of the DME into the pail.[90]
    a. For Ales, add 3-5 pounds (2 kg).
    b. For Lagers add 2-3 pounds (1 kg).
3) Add the distilled water to the fermenter.
4) Pour between to buckets to aerate and mix.
5) After two days add the remaining extract and the dry hops.

---

[85] http://www.briess.com/food/Processes/nstep.php
[86] http://www.muntonsmalt.com/about/how-malt-is-made/spraydrying/
[87] http://sciencefocus.com/qa/how-does-sugar-act-preservative
[88] http://en.wikipedia.org/wiki/Food_preservation
[89] See: "Yeast Rehydration"
[90] See "Easy Pitching"

Adding Dimension

A fine beer can be made with only extract, hops and yeast, but it may lack complexity. For a majority of beer drinkers this simple brew is expected. Personally, sometimes I even prefer something that is just simply beer. However many brewers want to quickly add more depth to their creation, and the fast brew method allows for this.

Specialty grains give the brewer detailed control over the flavor of their beer. These grains do not need to be mashed, or even boiled to impart their unique character to your brew. Ray Daniels' book "Designing Great Beers" is an excellent resource describing the characteristics of specialty grains. Half a pound of a specialty grain of your choosing is a good amount to start your experimentation.

1) While the yeast is rehydrating add the specialty grains to a pot with enough water to cover them. This is typically half a gallon per pound of grain.
2) Turn on the burner to medium high heat steeping the grains while stirring occasionally.
3) After 20 minutes, when the yeast has nearly completed hydrating, add the remaining water to the fermenter.
4) Pour the grains through a sanitized colander and into the ale-pail. If some grain gets into the beer do not worry.

There may be some concern of remaining bacteria after steeping. When the water temperature reaches 160°F (72°C) the grain is pasteurized killing a majority of the bacteria.[91] This is generally all that is required to make good beer. If the water boils for 15 minutes it is considered sterilized.[92] Keep an eye on the pot. Just a few seconds can make the difference between boiling and boiling over.

---

[91] http://en.wikipedia.org/wiki/Flash_pasteurization
[92] http://en.wikipedia.org/wiki/Sterilization_%28microbiology%29

Mix to Taste

Separating the beer into components opens up a pantheon of possibilities. Fermenting the beer with simply malt extract and yeast allows the hops and specialty grains to be added later. This allows the base beer to be tasted before the direction of the flavor is decided. After fermentation a small sample of the beer can be mixed. When working with metric units scaling the batch size down by a factor of one-thousand is convenient. For a 20-liter batch each taste would be 20 ml. This means, though, that some of the adjustments, especially hop tea, will require measuring very small volumes. Without a graduated 2 ml micro pipette this may prove to be difficult.

An alternative is to use US measurements. A five-gallon batch can be represented by a 3 tablespoon sample. At this scale an eighth of a teaspoon is equivalent to one cup at the batch scale. For example, you may find to get the mix you like, use a 3 tablespoon sample of beer adding 1 teaspoon of specialty grain tea and ⅛ teaspoon of hop tea. To duplicate this at the batch level you would need to add ½ gallon of specialty grain tea and 1 cup of hop tea to the 5-gallon batch. The table here should aid in conversations:

Batch Volume	Sample Volume
5 gallons	3 Tbls
1 gallon	2 tsp
1/2 gallon	1 tsp
1 quart	1/2 tsp
2 cups	1/4 tsp
1 cup	1/8 tsp

Table 44 - Batch to Sample Conversions

Recipes

15

Easy IPA

This recipe uses the "Fast Brew" method described in this book. In this beer there is enough hop bitterness to make it a solid IPA, but not so much that only hop heads will drink it. This is a favorite amongst my brewing friends, and is appreciated by many others as well!

Batch Size: 5 gallons
OG 1.060
FG 1.013
ABV 6.2%
Style: American IPA 14B
Type: Extract

Ingredients:
6 lbs Light Briess DME [93]
1 lbs sugar
2 cups of bittering hop tea.[94]
1 oz cirta hops for dry hopping[95]
1 pkg of US-05 Safale yeast[96]
5 gallons of distilled water (refrigerated)[97]

Start the hop tea, if you don't have some on hand, by adding 3 cups of water and 2 oz of Magnum hops to a sauce pan. This will simmer for one hour. Top the water off as needed to account for boil off.

Add 4.4 lbs of the DME and 3 quarts of distilled water to your boiling vessel and stir well to combine. Once the sugar is mostly dissolved turn on the heat and stir until all of the DME is dissolved.

---

[93] For hop centric beers Briess is preferred over Muntons. See "Great Beer with Little Time"
[94] See Bitter without the Boil
[95] For alternates see the "Low Alpha Hops" table
[96] See "Dry Yeast Comparison" table for alternates
[97] To create the correct pitching temperature the water must be refrigerated. See "Great Beer with Little Time"

While the pot is coming up to a boil clean and sanitize your fermentation bucket. Add the remaining 4 gallons and one quart of refrigerated distilled water to the fermenter.

Once the hop tea is ready and your wort has come to a boil add the ounce of Citra hops to the boiling wort. They will just be in the boiling wort for a few seconds. Pour the boiling wort into the cold water in the fermenter. Add one cup of hop tea to the fermenter. Aerate. Move your fermenter to your swamp cooler, or the location it will be fermenting. (Ideally the air temperature should be 60°-65°F in the location you are fermenting.) Sprinkle the yeast over the surface. The longer the yeast floats on the surface the higher the viability.[98]

Two days after pitching the yeast add the remaining 1.8 lbs of DME and the 1 lb of sugar to the fermenter. Allow this to ferment for two weeks. Fermentation will be complete in less than a week. The remaining time is for the beer to rest as the yeast clean up byproducts produced during fermentation.

Prime and bottle.[99] Allow at least one week to condition. Refrigerate for at least 48 hours.

Enjoy.

---

[98] See "Rehydrating Yeast"
[99] See "Easy Priming"

BREWING ENGINEERING

Brewing by the Numbers

Great beer often starts with a simple recipe that matches the style guidelines.[100] This table provides the vital recipe information to match most styles of beer. It can be a starting place to help decide what to brew next, or can be used as a finished recipe.

Style	DME Type	DME lbs/5 gal	Adj lbs/5 gal	Hop Tea	Yeast	Temp	Dry Hop
1A. Lite American Lager	Light	0.75	3.5 S	1/2 c	W 34/70	55°F	Hall
1B. Standard American Lager	Light	3.50	2.25 S	1/2 c	W 34/70	55°F	Hall
1C. Premium American Lager	Light	5.00	1.5 S	3/4 c	W 34/70	55°F	Hall
1D. Munich Helles	Munich	5.00	1 S	3/4 c	S-23	50°F	Hall
1E. Dortmunder Export	Munich	6.50		1 c	S-23	50°F	Hall
2A. German Pilsner (Pils)	Pilz	5.25	0.5 S	1 1/2 c	S-23	50°F	Hall
2B. Bohemian Pilsener	Pilz	5.50	0.5 MD	1 3/4 c	S-23	50°F	Saaz
2C. Classic American Pilsner	Pilz	6.25		1 1/4 c	W 34/70	55°F	Saaz
3A. Vienna Lager	Munich	6.00		1 c	S-23	50°F	Hall
3B. Oktoberfest/Märzen	Munich	6.25		1 c	S-23	50°F	Hall
4A. Dark American Lager	Dark	5.00	1.25 S	1/2 c	W 34/70	55°F	Tett
4B. Munich Dunkel	Dark	6.00		1 c	S-23	50°F	Hall
4C. Schwarzbier	Dk +1/4	6.00		1 1/4 c	S-23	50°F	Hall
5A. Maibock/Helles Bock	Amber	7.00	1.25 S	1 1/4 c	S-23	50°F	-
5B. Traditional Bock	Dark	7.75	0.5 S	1 c	S-23	50°F	-
5C. Doppelbock	Amber	9.00	1.25 S	1 c	S-23	50°F	-
5D. Eisbock	Amber	13.50	0.5 MD	1 1/4 c	S-23	50°F	-
6A. Cream Ale	Light	4.50	1.5 S	3/4 c	US-05	60°F	Lib
6B. Blonde Ale	Light	5.25	0.5 S	1 c	US-05	65°F	-
6C. Kölsch	Light	4.50	1.25 S	1 c	WB-06	65°F	-
6D. American Wheat	Wheat	5.00	0.75 S	1 c	US-05	60°F	Will
7A. North German Altbier	Dark	5.75		1 1/4 c	WB-06	65°F	-
7B. California Common Beer	Dark	6.00		1 1/2 c	US-05	60°F	NB
7C. Düsseldorf Altbier	Dark	5.75		1 3/4 c	WB-06	65°F	-
8A. Standard/Ordinary Bitter	Dark	4.25		1 1/4 c	S-04	70°F	Kent
8B. Special/Best/Premium Bitter	Dark	4.50	0.5 S	1 1/4 c	S-04	70°F	Kent
8C. Extra Special/Strong Bitter	Amber	6.50		1 3/4 c	S-04	70°F	Kent
9A. Scottish Light 60/-	Dark	3.00	0.5 MD	3/4 c	S-04	65°F	-
9B. Scottish Heavy 70/-	Dark	3.75	0.5 MD	3/4 c	S-04	65°F	-
9C. Scottish Export 80/-	Dark	5.25	0.25 MD	1 c	S-04	65°F	-
9D. Irish Red Ale	Dark	5.50	0.5 S	1 c	S-04	65°F	-
9E. Strong Scotch Ale	Dark	8.50	1.5 MD	1 c	S-04	65°F	-

Table 45 - Beer Styles 1-9

---

[100] http://www.bjcp.org/stylecenter.php

Style	DME Type	DME lbs/5 gal	Adj lbs/5 gal	Hop Tea	Yeast	Temp	Dry Hop
10A. American Pale Ale	Amber	6.00	0.25 S	1 1/2 c	US-05	70°F	Casc
10B. American Amber Ale	Dark	6.00	0.25 S	1 1/4 c	US-05	70°F	Casc
10C. American Brown Ale	Dk +1/4	6.25		1 1/4 c	US-05	70°F	Amar
11A. Mild	Dk +1/4	4.00	0.25 MD	3/4 c	S-04	70°F	-
11B. Southern English Brown	Dk +3/4	3.75	0.5 MD	3/4 c	S-04	65°F	-
11C. Northern English Brown	Dark	5.25	0.5 S	1 c	Nott.	65°F	Kent
12A. Brown Porter	Dk +1/2	5.50	0.25 S	1 c	Nott.	65°F	-
12B. Robust Porter	Dk +1/2	6.75		1 1/2 c	US-05	65°F	Kent
12C. Baltic Porter	Dark	9.00		1 1/4 c	S-23	55°F	-
13A. Dry Stout	Dk +3/4	4.50	1 S	1 1/2 c	S-04	65°F	-
13B. Sweet Stout	Dk +1	5.25	0.75 MD	1 1/4 c	S-04	70°F	-
13C. Oatmeal Stout	Dk +3/4	6.00		1 1/4 c	S-04	70°F	-
13D. Foreign Extra Stout	Dk +3/4	7.00	1.25 S	2 c	Nott.	65°F	-
13E. American Stout	Dk +3/4	7.25		2 1/4 c	US-05	65°F	Cent
13F. Imperial Stout	Dark	12.00		3 c	S-04	65°F	Kent
14A. English IPA	Amber	6.75	0.75 S	2 c	Nott.	70°F	Kent
14B. American IPA	Amber	6.75	1.25 S	2 1/4 c	US-05	65°F	Amar
14C. Imperial IPA	Amber	7.75	3 S	3 3/4 c	S-04	65°F	Cent
15A. Weizen/Weissbier	Wheat	6.00		1/2 c	WB-06	60°F	-
15B. Dunkelweizen	Wh +1/2	6.00		1/2 c	WB-06	60°F	-
15C. Weizenbock	Wh +1/2	8.75		1 c	WB-06	60°F	-
15D. Roggenbier	DK+1Rye	6.00	0.25 S	3/4 c	WB-06	60°F	-
16A. Witbier	Wheat	5.00	1 S	3/4 c	T-58	70°F	Coriander
16B. Belgian Pale Ale	Amber	5.75	0.25 S	1 c	T-58	65°F	Kent
16C. Saison	Amber	3.50	3.75 S	1 1/4 c	T-58	75°F	Hall
16D. Bière de Garde	Amber	6.00	2.75 S	1 c	T-58	65°F	-
17A. Berliner Weisse	Light	2.50	1.5 S	1/4 c	T-58	70°F	Hall
17B. Flanders Red Ale	Dark	3.50	3.25 S	3/4 c	S-04	65°F	-
17C. Flanders Brown Ale	Dark	5.00	2.25 S	1 c	S-04	65°F	-
18A. Belgian Blond Ale	Light	6.25	2 S	1 c	T-58	65°F	-
18B. Belgian Dubbel	Amber	6.25	2 S	3/4 c	T-58	65°F	-
18C. Belgian Tripel	Light	5.75	4.75 S	1 1/4 c	T-58	65°F	-
18D. Belgian Golden Strong	Ex Light	5.50	5.25 S	1 1/4 c	T-58	65°F	-
18E. Belgian Dark Strong	Amber	8.50	3 S	1 1/4 c	T-58	70°F	-
19A. Old Ale	Amber	9.00		2 c	Nott.	70°F	-
19B. English Barleywine	Amber	12.00		2 1/4 c	Nott.	70°F	Kent
19C. American Barleywine	Amber	11.25	1 S	3 1/2 c	US-05	70°F	Cent

Table 46 - Beer Styles 10-19

The "DME Type" Column in some cases contains the type of extract to use as well as additional steeping grains. The plus symbol followed by a number indicates the amount of chocolate malt in pounds to steep. For example "DK +1/4" indicates that dark dried malt extract should be used in addition to one quarter pound of chocolate malt. Other dark malts can be substituted for the chocolate to suit your pallet. Chocolate malt was chosen as it provides some roasted flavor without being overpowering.

The "DME lbs/5 gal" column indicates how much dried malt extract to use.

The "Adj lbs/5 gal" column shows the amount of adjuncts that will be used in a 5-gallon batch of beer. The letter code "S" indicates sugar, and "MD" is for Malto-Dextrin. The preceding number is the weight in pounds. For example "0.5 MD" is interpreted as half a pound of Malto-Dextrin.

The amount of "Hop Tea" is listed in teaspoons, tablespoons or cups. Instructions for preparing this tea can be found in an earlier section of this book.

One package of the yeast listed in the "Yeast" column should be used for the batch of beer. To ensure the proper pitch rate see the "Fast Brew Method" section of this book.

The intended fermentation temperature is listed in the "Temp" column.

For dry hopping, one ounce of the hop listed in the "Dry Hop" column can be used. These hops can also be added at flame out to ensure sanitation.

Raspberry Hefeweizen

Batch size: 2.5 gallons
OG: 1.053
FG: 1.010
ABV: 4.9
Style: 20 / 16A Fruit Witbier
Recipe type: All Grain

Grist:
2 lbs American 2 row
3 lbs Wheat Malt, Light
2 oz American Caramel L 40
8 oz Rice Hulls

Water Adjustments:
¼tsp CaSO4 (Gypsum) to bring the SO4 level to 65ppm
¼tsp CaCl (Calcium Chloride) to bring Cl level to 136ppm

Mash
Protein Rest at 122 for 15 minutes
Saccharification Rest at 152 for 60 minutes

Pitch
80 Billion cells of Safale - US-05 American Ale Yeast

Ferment
Swamp cooler at 68 degrees for one week
Pasteurize 4 lbs Frozen Raspberries, cool then add and ferment until final gravity is reached.

Bottle
Strain remaining fruit from the beer when racking to bottling bucket.
Prime with 3 oz of corn syrup dissolved in one cup of water.

Raspberry Cream Ale

The fall temperature is wonderful for a cream ale. It's not warm enough for an ale, and not cold enough for a lager. The cream ale gets its character from two primary contributors. The flaked corn gives it the creamy mouth feel, and the lower fermentation temperatures give it a refreshing cleanness normally associated with lagers. My wife and her friends love this beer so much that the only problem with it is that we can't make enough.

Batch size: 3 gallons
OG: 1.047
FG: 1.009
ABV: 5.0%
IBU: 15
Style: 20 Fruit Beer / 6A Cream Ale
Recipe type: All Grain

Grist
3 lbs 12oz (1700 grams) American 6 Row
4 oz (114 grams) Corn, Flaked
4 oz (102 grams) Cara Pils (Dextrine Malt)
3 oz (76 grams) Special B
5 oz (136 grams) Corn Sugar (added at flame out)

Water Adjustments
1/8 tsp $CaSO_4$ (Gypsum) to bring SO_4 level to 24ppm
1/4 tsp $CaCl$ (Calcium Chloride) to bring Cl level to 136ppm

Mash
Dough in at 122 and hold for at least 20 minutes while establishing a pH of 5.3 using lactic acid.
Saccharification rest for 1.5 hour at 149 degrees.

Hop Schedule

0.2 oz (6 grams) Challenger at 60 minutes (10.5 IBU)

0.15 oz (4 grams) Challenger at 10 minutes (2.9 IBU)

0.15 oz (4 grams) Challenger at 5 minutes (1.6 IBU)

Pitch

100 billion cells which is about one half package of Safale S-04.

Fermentation

63 degrees for 3 days, then 70 degrees for 4 days.
Rack to secondary. Add 3 lbs of pasteurized frozen raspberries.

Bottling

250 grams of corn syrup dissolved in 1 cup of water

(for corn sugar use 80 grams)

Comments

American 6 row is the traditional grain used for this style although many recipes use 2 row.

Flaked Corn adds creaminess. Flaked corn was chosen over other forms because it does not need to be cooked before the mash.

CaraPils adds body and aids in head retention.

Special-B is used in place of Caramel because of its cake like flavor instead of taffy flavor.

Corn Sugar will lighten the body of the beer. Using both corn sugar and carapils in the same recipe may seem like a conflict, but the result is the best of both worlds: the head retention of the carapils and the lightness of the sugar.

For ale, the temperature is in the lower range for primary fermentation. Bringing the temperature up to 70 degrees after most of the fermentation is complete allows the yeast to clean up byproducts.

Hoegaarden Clone Recipe

Batch size: 5 gallons
OG: 1.051
FG: 1.014
ABV: 4.9
Style: 16A Witbier
Recipe type: All Grain

Grist:
4 lbs 12oz Belgian Pilsner
4 lbs 12oz Wheat, Light, Malted
1lb Dextrine Malt (Carapils)

Water Adjustments
1/8 tsp $CaSO_4$ (Gypsum) to bring SO_4 level to 24ppm
½ tsp $CaCl$ (Calcium Chloride) to bring Cl level to 136ppm

Mash
1.5 hour ramp from hot tap water to sparge temperature.

Boil Schedule
1.75 oz crushed (not ground) coriander at 60 minutes
1 oz Glacier at 20 minutes (15 IBU)
1 oz Glacier at flame out (1 IBU)
1 oz of sweet orange peel at flame out

Pitch
180 billion yeast cells. Prepare a 6 cup starter (of 1.040 wort) with a smack pack of 3056 Bavarian Wheat Blend Yeast, or use two smack packs.

Fermentation
64-74 degrees recommended for yeast.

Hold at 64 degrees for one week in a swamp cooler then move to 70 degree ambient air environment for the remainder of fermentation.

Bottling
3oz of corn syrup (or 1 oz of table sugar) dissolved in 1 cup of water[101]

Notes

The recipe used today contains 55% Barley Malt, and 45% Wheat.[102] The FG is about 4 points higher that would be expected given typical 75% attenuation, so 10% of the grist was changed to a dextrin malt. CaraPils was chosen for its light color.

The water in most places in the US is calcium deficient so only calcium salts are used. The amount will vary dependent on your water. Adjust as needed to achieve the final levels.

The Hoegaarden website states a 2 hour gradual rise to 170. With my BIAB stove top setup this will be nearly full heat in the 4 gallon pot that I use to achieve this ramp rate.

Coriander does pretty well in the boil from what I have read in "The Home Brewers Garden".

The low end of the fermentation range was chosen to aid with the higher than normal final gravity of this beer. After primary fermentation is compete -- in about a week -- moving the beer to a warmer environment will allow the yeast to clean up any off flavors that may have been created during the primary fermentation.

---

[101] http://woodlandbrew.blogspot.com/2012/10/priming-with-corn-syrup.html
[102] http://www.whitebeertravels.co.uk/celis.html

Equations

16

Overview

Sugar (g)	100	87.5	75	62.5	50	37.5	25	12.5	0
Yeast Grown (g)	0.0	0.7	1.3	2.0	2.7	3.3	4.0	4.7	5.3
Yeast (billion of cells)	0.0	13.3	26.6	39.9	53.2	66.5	79.8	93.2	106.5
Water (ml)	1000	1000	1000	1000	1000	1000	1000	1000	1000
Alcohol (ml)	0	8	15	23	31	38	46	54	61
Refractometer	9.1°B	8.1°B	7.1°B	6.2°B	5.2°B	4.2°B	3.1°B	2.1°B	1.1°B
Sugar by Weight	9.1%	8.0%	6.9%	5.8%	4.7%	3.5%	2.4%	1.2%	0.0%
ABV	0.0%	0.7%	1.4%	2.2%	2.9%	3.6%	4.3%	5.1%	5.8%
Specifig Gravity	1.031	1.026	1.020	1.014	1.009	1.003	0.997	0.992	0.986
Apparent attnuation	0%	18%	36%	54%	72%	90%	108%	126%	144%
Actual Attenuation	0%	13%	25%	38%	50%	63%	75%	88%	100%

Sugar (g)	200	175	150	125	100	75	50	25	0
Yeast Grown (g)	0.0	1.3	2.7	4.0	5.3	6.7	8.0	9.3	10.6
Yeast (billion of cells)	0.0	26.6	53.2	79.8	106.5	133.1	159.7	186.3	212.9
Water (ml)	1000	1000	1000	1000	1000	1000	1000	1000	1000
Alcohol (ml)	0	15	31	46	61	77	92	107	123
Refractometer	16.7°B	14.9°B	13.2°B	11.4°B	9.6°B	7.7°B	5.9°B	4.0°B	2.1°B
Sugar by Weight	16.7%	14.8%	12.8%	10.8%	8.7%	6.6%	4.5%	2.3%	0.0%
ABV	0.0%	1.4%	2.7%	4.1%	5.4%	6.8%	8.2%	9.6%	10.9%
Specifig Gravity	1.060	1.050	1.039	1.029	1.018	1.007	0.997	0.986	0.975
Apparent attnuation	0%	17%	35%	53%	70%	88%	105%	123%	141%
Actual Attenuation	0%	13%	25%	38%	50%	63%	75%	88%	100%

Table 47 - Beer Constitues Durring Normal Fermentation

The information in this table may be best visualized by thinking of this as two beers of different gravity. Each column represents a snap shot in time from inoculation to fermentation of all sugars. Note that doubling the fermentable sugar doubles all of the products. These are alcohol, yeast and carbon dioxide. Because of the difference in density and index of refraction between water and alcohol, the Specific Gravity measurements and refractometer measurements are not simply multiplied.

Equations that are "based on a physical model" are algebraically derived from other equations. These include the Balling Observation that 2.0665 g of sucrose are required to create 1 g of Ethanol. They also use some basic analytical chemistry to convert from various units.

Equations labeled as "extrapolated from a physical model" are derived from the physical models as describe above. Instead of listing complicated equations that are algebraically correct these equations are simplified linear fits to the physical equations. In most brewing applications these should be just as accurate as their algebraic equivalents. When compared to the physical model these linear fit equations were accurate to two decimal places, or 1%.

Lautering

When adjusting a recipe, or formulating a new one, a firm grasp of lautering is essential.

Find: *Efficiency (E)*

From: Water added to the mash (W) –and–
Grain weight in pounds (G)

$E = (W-0.15G)/W$
   Ex: (6-0.15(10))/6=0.75 (or 75%)

Wort Properties

Find: **Percent Sugar by Weight (°P)**

From: Specific Gravity (SG)
SBW = 250(SG-1)
   Ex: 250(1.060-1)=15°P

Find: *Specific Gravity*

From: Percent Sugar By Weight (°P)
SG = (SBW/250)+1
   Ex: (15/250)+1=1.060

BREWING ENGINEERING

Beer Properties

After fermentation these equations can be used to derive some of the key properties of the beer.

Find: *Alcohol by Volume (ABV)*

From: Final Refractometer measurement in Brix (FR) –and–
Final Hydrometer measurement in specific gravity (FH)

ABV =1.081*FR-0.273*FH-0.053 (accurate to 1% for most beers)
Ex: 1.081(9.6)-0.273(18)-0.053=5.4 (or 5.4% ABV)

From: Original Hydrometer in specific gravity (OH) –and–
Final Hydrometer measurement in specific gravity (FH)

ABV =131(OH-FH)
Ex: 131(1.060-1.008)=6.8 (or 6.8%)

From: Original Refractometer in degrees Brix (OR) –and–
Final Refractometer measurement (FR)

ABV =0.6967*(OR-FR) (extrapolated from physical model)

Ex: 0.6967(12-5.4)=4.6 (or 4.6%)

Find: *Percent Sugar By Weight (SBW)*

From: Original Refractometer measurement in Brix (OR) –and–
Final Refractometer measurement in Brix (FR)

SBW =1.133*FR-0.1389*OR+0.039 (extrapolated from physical model)

*Ex: 1.133*7.7-0.1389*16.6+0.039=6.5 (or 6.5%)*

From: Final Hydrometer measurement in Brix (FH) –and–
Final Refractometer measurement in Brix (FR)

SBW =0.7488FR+68.55*FH-68.60 (extrapolated from physical model)

Ex: 0.7488(7.3)+68.55(1.018)-68.60= 6.7 (or 6.7%)

Find: ***Original Gravity (OG)***

From: Final Hydrometer measurement in Specific Gravity (FH) –and–
Final Refractometer measurement in Brix (FR)

OGH =-1.728*FH+0.01085*FR + 2.728 *(extrapolated from physical model)*

Ex: -1.728*1.007+0.01085*7.7 + 2.728=1.071

Find: ***Final Gravity (FG)***

From: Original Hydrometer measurement in SG (OH) –and–
Final Refractometer measurement in Brix (FR)

FG =0.00628*FR-0.0025*OR+1.0013 *(based on physical model)*

Ex: 0.00628*8.8-0.0025*15.2+1.0013=

Attenuation

Apparent Attenuation = 1-(OG/FG)

Actual Attenuation = 0.71(AA)

(3% accuracy for beer with an OG of 1.020 to 1.100)

Ex: 0.71(78) =55 (or 55%)

Yeast Growth

Cells Grown (Billions) = 14 * Volume of wort (Liters) * [Initial Gravity of wort (°P) - Final Gravity of wort (°P)]

Index

acid, 34, 124
acid washing, 125
acidic acid, 181
aerobic respiration, 78, 79, 80, 94
alcohol tolerance, 118
anaerobic respiration, 78
attenuation, 12, 24, 28, 46, 56, 67, 85, 126
bacteria, 56, 62, 104, 109, 110, 116, 123, 124, 125, 126, 148, 149, 156, 158, 164, 167, 170, 186, 200, 201, 203, 206, 207
Balling Observation, 79, 87, 88, 173, 222
BIAB, 38, 54, 220
bitterness, 12, 64, 191, 201, 202
Briess, 55, 197, 198
BU GU, 58, 60
buffer, 34
calibrate, 40, 41
Campden tablets, 32, 33, 189
Carapils, 19
carbonation, 132
cell count, 180
cell density, 105, 171
cell growth, 85
cellulose, 62, 111
chloramine, 32
chlorine, 32
cooling wort, 100
Coopers, 198
cost, 124, 164, 195, 202
crystal malt, 12, 19, 25, 27
density, 153
dextrin, 12, 23, 24

diacetyl rest, 128
dry malt extract, 25, 56, 164, 195, 200, 201
dry yeast, 93, 204
efficiency, 38, 39, 48, 49, 54
enzymes, 55, 57, 189
ester, 14, 102
fermentability, 24, 25, 27, 39, 189, 197
fermentation vessel, 99, 105, 118, 173, 205
final gravity, 12, 23, 24, 27, 57, 92, 128, 130, 188, 225
fruit, 62, 112, 121, 191, 215, 216
gelatin, 172
glycogen, 70, 84, 85
grain absorption, 51
heater, 142
hemocytometer, 158, 180, 182, 183
Hoegaarden, 17, 219, 220
hop tea, 130, 133, 191, 192, 202, 208, 210, 211
hop utilization, 201
hops, 20, 60, 64, 128, 191
hydration, 93
hydrometer, 153, 173
inoculation rate, 72, 82
ion, 12, 13, 30, 34, 35, 125, 199
lactic acid, 34, 35, 123, 216
lactose, 26, 27
lag phase, 70
lautering, 38, 48, 223
mash, 46, 53
mash pH, 34
mash time, 25, 47, 189

mash tun, 38, 48, 51, 53, 58
methylene blue, 70, 118, 181, 182
microscope, 72, 91, 98, 100, 113, 121, 156, 164
microscope focus, 156, 161, 184
mold, 119
mouth feel, 12, 14, 216
Muntons, 198
oxygen, 57, 78, 88, 94, 102
pasteurize, 148
petri dish, 167
phenol, 102
phenolic, 14
phenols, 142
pipette, 160
pitch rate, 69, 97
plate, gelatin, 167
plate, gelatin, 171
preservative, 201, 203, 206
Ray Daniels, 58, 64, 188, 207
refraction index, 153, 173, 175, 176, 188
refractometer, 153, 173
Saison, 24, 91, 122, 143
salt, 13, 31, 126
soy sauce, 162
sparge, 38, 48, 49, 52, 54
specialty grain, 192, 207, 208
SRM, 18, 162
Star San, 125, 126
starch, 39, 48, 70, 189
sterilize, 147, 167, 207
sulfur dioxide, 33
swamp cooler, 136, 142, 144
sweetness, 12, 19, 23, 30, 191
taste, 12, 58, 65, 142, 191, 192, 208
temperature, 39, 48
test tubes, 161
trub loss, 133, 195
viability, 68, 70, 100, 112, 114, 116, 126
vitality, 70
water, 30
White Labs, 18, 73, 95, 98, 182, 186
yeast propagation, 104
yeast rehydration, 94
yeast washing, 109, 123

Printed in Poland
by Amazon Fulfillment
Poland Sp. z o.o., Wrocław